Joseph W. Jerome Analysis of Charge Transport

Springer
Berlin
Heidelberg
New York
Barcelona
Budapest
Hong Kong
London
Milan
Paris
Santa Clara
Singapore
Tokyo

Joseph W. Jerome

Analysis of Charge Transport

A Mathematical Study of Semiconductor Devices

Springer

Joseph W. Jerome
Northwestern University
Department of Mathematics
Evanston, IL 60208, USA
e-mail: jwj@math.nwu.edu

Cataloging-in-Publication Data applied for

Die Deutsche Bibliothek - CIP-Einheitsaufnahme

Jerome, Joseph W.:
Analysis of charge transport : a mathematical study of semiconductor devices / Joseph W. Jerome. - Berlin ; Heidelberg ; New York ; Barcelona ; Budapest ; Hong Kong ; London ; Milan ; Paris ; Santa Clara ; Singapore ; Tokyo : Springer, 1996
ISBN-13:978-3-642-79989-1

Mathematics Subject Classification (1991): 35J50; 35J70; 41A46; 65L70; 65M60; 82B21; 82C70; 82D25

ISBN-13:978-3-642-79989-1 e-ISBN-13:978-3-642-79987-7
DOI: 10.1007/978-3-642-79987-7

Softcover reprint of the hardcover 1st edition 1996

Typesetting: Camera-ready copy produced from the author's output file using a Springer TEX macro package
SPIN 10500125 41/3143-5 4 3 2 1 0 - Printed on acid-free paper

For My Children

Jon

* March 30, 1968
† April 27, 1992
Beloved, as in life

Peter

Son and dear friend

The deeper that sorrow carves
into your being, the more joy
you can contain.

Kahlil Gibran

Preface

This book addresses the mathematical aspects of semiconductor modeling, with particular attention focused on the drift-diffusion model. The aim is to provide a rigorous basis for those models which are actually employed in practice, and to analyze the approximation properties of discretization procedures.

The book is intended for applied and computational mathematicians, and for mathematically literate engineers, who wish to gain an understanding of the mathematical framework that is pertinent to device modeling. The latter audience will welcome the introduction of hydrodynamic and energy transport models in Chap. 3.

Solutions of the nonlinear steady-state systems are analyzed as the fixed points of a mapping $\mathbf{T}$, or better, a family of such mappings, distinguished by system decoupling. Significant attention is paid to questions related to the mathematical properties of this mapping, termed the Gummel map. Computational aspects of this fixed point mapping for analysis of discretizations are discussed as well.

We present a novel nonlinear approximation theory, termed the Krasnosel'skii operator calculus, which we develop in Chap. 6 as an appropriate extension of the Babuška-Aziz inf-sup *linear* saddle point theory. It is shown in Chap. 5 how this applies to the semiconductor model. We also present in Chap. 4 a thorough study of various realizations of the Gummel map, which includes non-uniformly elliptic systems and variational inequalities. In Chap. 7 we present the fundamental difference between basing Newton's method on the fixed point map, as distinct from the differential formulation. The latter requires smoothing, while the former serves as its own smoother. We are led in this chapter to formulate the central approximation problem of drift-diffusion theory, and to describe its solution.

The drift-diffusion model is still a mainstay in the engineering community for device simulation, and it finds utility more generally in various electro-diffusion processes in electro-chemistry and biophysics, including the study of channels in biological membranes. We have selected a wide range of topics to present, some completely new, which we believe cover practical as well as theoretical issues of import. In particular, Chaps. 2, 3, and Sect. 4.1 should be accessible to most engineers, while mathematically inclined scientists will find Chaps. 4, 6, and 7 of interest. The finite element community may find Chaps. 5 and 6 useful, while applied mathematicians may find particular interest in Chaps. 2, 3, and 5.

It is my pleasure to acknowledge my appreciation of the ongoing relationship which I have had with Thomas Kerkhoven, whose perspective has influenced this work, and who assisted with part of it, particularly some results of Chap. 5. I would also like to thank C. T. Kelley for the observation which led to the extension, contained in Chaps. 5 and 6, of the original nonlinear approximation calculus developed by the Russian school. This observation dealt with the breakdown of Lipschitz continuity of the derivative of the truncation map, and necessitated the construction of a theory based upon a joint energy and pointwise norm. Carl Gardner and Thomas Seidman read versions of the manuscript, and made many suggestions for improvement. I am grateful for their assistance. A visit to the University of Maryland in 1987, and ensuing conversations with Ivo Babuška and John Osborn, proved strongly motivating in understanding the interface between the linear and nonlinear theory. Finally, I should like to express my appreciation to my good friend, Richard Varga, for his encouragement to pursue this project.

Table of Contents

Part II. Computational Foundations

1. Introduction

The general themes of this book attempt to capture timely issues of modeling, theoretical computation, and mathematical theory. Engineers, scientists, and mathematicians now accept nonlinear systems of the scope developed here as basic for the analysis and prediction of the processes of charge transport, in a self-consistently determined electric field. In this brief introductory chapter, we shall expand upon the remarks of the preface, and discuss the goals of the book, and their development. Some historical detail will be furnished, insofar as it relates to the thematic aims.

The present climate is one of encouragement toward cooperation among academic, industrial, and government laboratory scientists. In 1980, when the author first became involved with the mathematical modeling of semiconductor devices, there was far less definition in this respect. At that time, perhaps the most significant immediate example of such feedback involved the petroleum industry, particularly reservoir simulation. The systems of partial differential equations studied in reservoir simulation had occupied a number of academic and industrial scientists, often in consultation. They are not unlike those of this monograph, viz., self-consistent drift-diffusion systems. There, however, the most important problem is the time dependent modeling.

One of the major centers for basic microelectronic research in the early 1980s was AT&T Bell Laboratories in Murray Hill, New Jersey. After spending the summer of 1981 at Murray Hill, at the invitation of Dr. James Mc Kenna, the author arranged to spend the academic year 1982–83 as well, at the invitation of Dr. Donald Rose and Dr. Norman Schryer. The simulation group at Bell Laboratories had developed a cutting edge two-dimensional algorithm for the steady-state model, based upon a damped Newton outer operator iteration, in conjunction with Scharfetter–Gummel operator discretization, and an inner relaxation or conjugate residual iteration to solve the linear algebraic systems. A considerable amount of the author's mathematical effort during the balance of the decade was an attempt to provide the mathematical foundation for the success of this algorithm, and its preliminary smoother, so-called Gummel iteration. This monograph has emerged from these studies and interactions with individuals from Bell Laboratories and elsewhere. In the middle 1980s, largely through the efforts of the late Farouk Odeh, at the IBM T.J. Watson Research Center, the hydrodynamic model emerged as a viable modeling tool. The author was in active dialogue with Odeh at that time, and was no doubt influenced by these developments. Concurrently, Stanley Osher of UCLA had developed a powerful shock capturing code (ENO) with his collaborators and students, and the relationships developed by the author with this group of scientists were to prove decisive in introducing this algorithm to semiconductor simulation. A final comment is worth noting. The drift-diffusion model, under the nomenclature of the Poisson-Nernst-Planck model, has achieved recent acclaim in the biophysical community, particularly under the leadership of Dr. Robert Eisenberg of Rush Medical College, where it has shown exceptional robustness in fitting experimental data for the open ionic channel.

The monograph has been organized into three parts; on modeling, computational foundations, and mathematical theory, respectively. In the remainder of the introduction, we shall give the reader an excursion through these parts, so that their rationale is clear.

1.1 Modeling

This part of the monograph, comprising the first two chapters, establishes a hierarchy of models derived from moments of the Boltzmann equation. These are all self-consistent models in the sense that the electrostatic potential is not predetermined or given, but is established by the transport processes, and hence the free carrier concentrations, the doping, and the contact boundary conditions. In Chap. 3, we develop both the hydrodynamic model, a perturbed conservation law system, built up from charge, momentum and energy exchange, and an energy transport system, which differs from the hydrodynamic model, in that it lacks hyperbolic modes. Both models attempt to deal adequately with so-called hot electron effects, including carrier velocity overshoot. In this context, we present a variety of one-dimensional results, including a study of the linearized hydrodynamic model, especially in the subsonic (elliptic) case, and the well-posedness of the energy transport model. The application for these analytical results is the n^+-n-n^+ diode. We are not aware of any well-posedness (existence) results for either of these models in two or more dimensions, except in certain cases of radial symmetry. We also present the outline of a new model for quantum fluid transport, which has recently been introduced, and for which rigorous results are available for a simplified case of the model.

In Chap. 2, we present various simplified versions of the drift-diffusion system. In the process, we explain how the Scharfetter–Gummel discretization is interpreted in terms of generalized splines: a piecewise constant flux method. As an explicit example of the drift-diffusion system, we solve approximately for the equilibrium configuration in a p/n junction, including the width of the boundary layer. In addition, in this chapter, we describe the critical role played by mobility functions in the drift-diffusion model.

The functional form derived in this chapter is precisely attuned to the saturation property of current versus electric field, a property observed in real transistors. Finally, we introduce in Chap. 2 the quasi-Fermi levels, obtained from the electron and hole concentrations by exponential transformations. When the system boundary-value problem, mixed in character, is formulated in terms of these variables, one finds that the drift and diffusion components fuse into a 'de facto' diffusion formulation, if the Einstein relations are employed. In simulations, the Fermi levels are often the variables of choice because of their more limited numerical range. Although the book does not deal with simulation per se, references are given in Chap. 3 to the recent computational literature on the hydrodynamic and energy transport model, and, in Chap. 2, major simulators are cited for the drift-diffusion model.

1.2 Computational Foundations

When the author formulated an existence proof for the drift-diffusion model in 1980, shortly after a visit to Yale University as a guest of Martin Schultz, the method of choice was the construction of a fixed point map, which proceeded by operator decoupling. Such mappings are now organized under the (categorical) name of the Gummel map. Iteration with this mapping, when convergent, is a way of computing the solution approximately, via the numerical solution of individual gradient equations, if complete system decoupling is selected. All effective numerical methods make use of this iteration as *part* of their strategy! It is important to emphasize that the Gummel map is important even when these iterates fail to converge. An existence result proceeds from the Schauder fixed point theorem in this case. Moreover, as we show in this monograph, the mapping, and a companion fixed point mapping, can be used to describe the convergence of Galerkin approximations, and, even more significantly, can be used to define linear approximation problems, based upon Newton's method applied to the numerical fixed point map, for the construction of convergent optimal order approximations.

In Chap. 4, we carry out an exhaustive study of the family of Gummel mappings, while focusing on admissible decouplings of the current continuity subsystem, which lead to individual members of this family. An early result in this chapter computes the Lipschitz constant for the solution map corresponding to the two-dimensional mixed boundary-value problem. If this constant is strictly less than unity, then the fixed point mapping is strictly contractive, and the successive approximation iterates converge. The Lipschitz constant is proportional to the energy band bending and the quantity $d^{5/2}$, where d is the device diameter. The result assumes proven gradient singularities at the boundary transition points of order $r^{-(1/2)}$. This is the only result of the book which depends fundamentally upon the Slotboom variables, which appear in a few other instances as convenient variables for estimation. Uniqueness for the device equations follows when the Lipschitz constant is less than one, but not otherwise. A major open problem in this subject is uniqueness in general. Even in one dimension, the issue is very subtle. For example, uniqueness does not hold in the case of the thyristor, which possesses three p/n junctions. On the other hand, it is known to hold in the absence of doping under charge neutrality at the boundary.

Since effective computations are frequently based on the Gummel map, we have studied the composition mappings, which define this map, in detail. When the Fermi levels are employed, a member of the family can be written as $\mathbf{T}_f = \mathbf{VW}_f \circ \mathbf{U}_f$. Here, the mapping $\mathbf{U}_f$ is obtained from solution of the potential equation for prescribed values of the other dependent variables. The map designated $\mathbf{VW}_f$, to stress the Cartesian product image space for the range of the map, solves the (partially) decoupled current continuity subsystem. Those decouplings which lead to a well-defined mapping are called admissible; they are based on lagging of the recombination term. The well-posedness of this map is carried out via variational inequalities, in order to develop a framework in which the maximum principles are verifiable. They are incorporated as "obstacles", and are later shown to be "nonbinding", so that the variational inequality is equivalent to a system of equations. A noteworthy aspect of this analysis is that the current continuity subsystem is degenerate elliptic, and the use of weighted spaces is required. This technical issue arises because, according to Einstein's relations, the mobility and diffusion coefficients are proportional at constant temperature. However, the mobility coefficients employed, to model saturation, tend to zero at boundary transition points, where the electric field modulus is unbounded. This leads to the breakdown of uniform ellipticity, which is analyzed. Such an analysis is not available elsewhere in the semiconductor literature to the author's knowledge.

In Chap. 5, we develop a calculus for the composition mappings of the system, and the corresponding numerical composition mappings, based on piecewise linear finite elements. The purpose of this calculus, when combined with an approximation theory for the mappings, is to provide a Galerkin approximation theory for the finite element method. This is possible when the Galerkin approximations are viewed as fixed points of the numerical fixed point mappings. Discrete maximum principles prove pivotal here. In this chapter, we assume constant mobility and diffusion coefficients, and zero recombination, so that certain technical issues related to the composition mappings are not present. To the author's knowledge, these results, when coupled to the theory of Chap. 6, give the first comprehensive steady-state Galerkin approximation theory for self-consistent drift-diffusion systems. Note that linearization is a fundamental component of this theory.

1.3 Mathematical Theory

Chap. 6 provides a functional analysis framework, via a calculus and convergence estimates due to Krasnosel'skii and his collaborators, for the explicit calculus developed in the preceding chapter. We present this in considerable detail, including a necessary generalization involving a second (stronger) norm, required for the application. As part of the discussion, we show how the Babuška-Aziz linear saddle point theory is subsumed under the aegis of the nonlinear estimation theory. One of the significant results of this chapter, which could be overlooked, deals with the replacement of the exact numerical fixed point (read Galerkin approximation for the application of the preceding chapter) with an approximate fixed point, constructed from linearization of the numerical fixed point map. It happens that, for decreasing approximation errors, the local linear approximation projected onto the nonlinear problem becomes progressively more accurate. The reader might be inclined, incorrectly, to view this as simply a refinement of the Galerkin approximation results. This result actually represents a major step as elaborated in the following Chap. 7. Since the linearization is based upon the numerical fixed point map, it is not computed by linearizing the Galerkin equations. It has been detached from the differential system directly, and thereby retains the spirit of Gummel iteration in terms of Newton approximations. We explain in Chap. 7 how this is essential if one wishes to construct an approximation sequence which is optimal, computable, and stable, and thereby avoids the numerical loss of derivatives inherent in basing Newton's method upon the differential system. The latter leads to constants in Newton estimates which are mesh dependent. In Chap. 7, we also present the numerical analysis version of the Nash-Moser iteration theory, which, to the author's knowledge, is the only rigorous antidote to the numerical loss of derivatives encountered in basing an approximate Newton method upon a standard numerical method, when the (noncompact) differential map is linearized. This is presented for general interest only; it is not involved in the constructions presented here, since we use compact mappings as distinct from noncompact mappings. Chap. 7 also contains a statement of what we characterize as the central approximation problem for self-consistent drift-diffusion systems, as well as the sense in which this problem is resolved.

1.4 Summary

Chap. 3 presents a diverse class of device models, cutting across the classifications of partial differential equations, for semi-classical and quantum transport of charged carriers. A number of topical papers dealing with simulation are referenced in this chapter, and elements of theory are developed. The attempt is made to reconcile microscopic and macroscopic approaches.

A noteworthy feature of the monograph is the solution of the mathematical problem of constructing a nonlinear finite element convergence theory, carried out in Chap. 5. However, the routine modular thinking, of replacing the resulting standard finite dimensional nonlinear formulation, by a sequential Newton method, in conjunction with matrix iterative methods, is shown to be inadequate. Rather, the interface with the nonlinear finite element problem is identified as critical from the standpoint of computational complexity. A proper iteration is defined, in a computable manner, in relation to the (numerical) fixed point mapping. Chaps. 6 and 7 deal with the corresponding issues. It is in this sense that we have established a mathematical foundation for algorithms similar to, and yet modifications of, the successful algorithms first developed more than a decade ago.

Finally, although the existence theory for drift-diffusion models is an important consequence of Chap. 4, more important is the study of the system mappings. Major technical problems are overcome by the interpretation of maximum principles as constraints, together with the use of variational inequalities. The nonbinding character of the constraints is a beautiful feature of these systems, leading to the equivalence with the current continuity subsystem.

Part I

Modeling of Semiconductor Devices

2. Development of Drift-Diffusion Models

2.1 Descriptive Background

The basic principle underlying the function of a transistor element is that current flows between oppositely directed p/n junctions (or does not flow), depending upon controlling currents or voltages elsewhere in the element. A junction occurs when regions with contrasting doping characteristics abut in a semiconductor. Thus, a p-region is one with an excess of free hole carriers and an n-region, on the other hand, contains an excess of free electrons. Such charge imbalances are induced by the process of doping, whereby impurity concentrations are injected into pure crystalline semiconductor materials. From the standpoint of valence chemistry, the impurity atoms have numbers of electrons in their outer shells different from the semiconductor atoms. Preponderance of donor impurities creates n-regions as well as net concentrations of positive ions, and acceptor impurities create p-regions and negative ions. The net charge at any point in the device is obtained by combining the ions with the free electron and hole carriers. Thus, an electric field is created via the first Maxwell equation, often called the Poisson equation in the theory of electrostatics. There are also quantum mechanical interpretations related to the impurity injection process; these include energy band bending and wave packet identifications as classical particles (see [87, 122]).

A transistor is a three terminal device, and this distinguishes it from a diode which is a two terminal device. Although there are various designs for transistors, the two fundamental categories are those of bipolar junction, and field effect transistor. In the bipolar junction transistor (BJT), consisting of emitter, base, and collector regions with respective terminals, small variations of the current in the base region lead to significant changes in the collector current; this property is known as amplification. These current variations can also switch the device, i.e., take it from the "on" or saturation state, where a large value of collector current is achieved via a small voltage drop from emitter to collector, to the "off" or cutoff state, where the base current is zero or negative. The field effect transistor (FET) is controlled by a voltage at the third terminal, called the gate, rather than by a current. In this case, carriers flow from source contact area to drain contact. This type of transistor is sometimes called a unipolar device because the current is dominated by majority carriers. The two principal types of FET devices are the junction

FET device, which normally possesses a conducting channel until reverse bias on the gate induces pinch-off, and the MOS-FET device, for which a conducting channel is created by application of a positive voltage at the gate. The MES-FET is sometimes included too, although it functions similarly to the junction FET device. See [130] for further discussion. See also the recent text of Lundstrom [99].

In all cases, the information required by the device physicist includes the current-voltage response curves, or the derivatives of these curves, which allow the device to be modeled as part of an integrated circuit. Sometimes the number of devices is small, such as when one wishes to simulate a single gate in a logic circuit, and sometimes quite large, as when one is attempting the electrical or timing response of an entire circuit (for the latter, see [136]). The role of device modeling, then, fits in as the interface between process and circuit modeling. The device geometry and the doping profile are the desired output of process modeling, which, in the context of free boundary problems, allows for diffusive, thermal, elastic, viscous, and oxidation effects (see [38]). Geometry and the doping profile then become the input of device modeling. At the present time, inaccurate process output constitutes, perhaps, the major barrier to a far reaching CAD implementation of device simulation. It is the purpose of this monograph to analyze device modeling in detail, and to develop results associated with the mathematical and approximation theoretic aspects of the models. Neither the number of terminals nor the number of junctions is critical to most of our investigations. Moreover, junctions where the doping magnitude changes, but not the sign, are also of interest. Thus, the well-known n^+-n-n^+ diode is within the scope of this investigation; the latter attempts to simulate the channel of a MOS-FET transistor.

It was almost fifty years ago, in 1947 to be precise, that the transistor was successfully developed at Bell Laboratories in Murray Hill, New Jersey, by John Bardeen, Walter Brattain, and William Shockley, an achievement for which they were awarded the Nobel Prize in 1956. The consequences of this discovery have not yet been fully realized, and this monograph is a recognition of the ongoing studies in this area.

2.2 Modeling Overview

It is advantageous to view classical or semiclassical semiconductor modeling as an hierarchical structure, in which the Boltzmann transport equation forms the summit, and the drift-diffusion model the base; intermediate are systems derived from moments of the Boltzmann transport equation. The hydrodynamic model and its inertial limit, a type of energy transport model, are examples. Presently, there exist two fundamental approaches to obtaining solutions of the Boltzmann transport equation, viz., Monte-Carlo (statistical sampling) methods and deterministic methods. The latter are of spectral type, applied to momentum space, or are of the recently developed particle type. These methods tend to be costly, irrespective of their type. As many as three orders of computational complexity separate the Boltzmann simulations from those beneath it in the hierarchical structure. See [39, 59, 135] for the Monte-Carlo method, [95] for the spectral method, and [32, 107] for the particle method. The hydrodynamic model offers the promise of simulation detail, as well as feasibility (see §3.1 for a derivation). It does have the drawback of requiring adequate closure assumptions and accurate representations for the average collision mechanisms. In addition, the model has hyperbolic as well as parabolic components for the moment equations, in the transient case, and thus issues of shock capturing mechanisms arise. On the other hand, the energy transport model, allowing for carrier heating, and the drift-diffusion model, the constant temperature case of the energy transport model, have only parabolic components, exclusive of the Poisson equation, which must be adjoined to any model in the hierarchy.

An interesting derivation of the drift-diffusion model from the Boltzmann transport equation, making use of the diffusion approximation, is carried out in [31]. This model is by far the best understood of the above models, and it is the primary topic of this monograph. It was introduced by Van Roosbroeck in 1950 [113] as a conservation system for electron and hole carriers, with the ambient electric field determined from the Poisson equation. This model is closely related to the older (by sixty years) ionic transport model of Nernst and Planck, which is described in detail in [115]. Although the analogies are fascinating, we shall not have occasion to draw upon them in this work. It is of considerable interest, however, that the current-voltage curves in biological membrane channels are derived from mechanisms closely related to those described here, with similar systems of equations involved in the modeling.

When appropriate scalings are selected for the quantities appearing, it is found that the conservation part of the system is highly convective. A simple illustrative case is examined in the next section to indicate this. There is a scaling, viz., that used to bring the order of the electron and hole concentrations to unity, in which the convection domination is associated with a potential, arising from a Poisson equation, which can be singularly perturbed. The combination of these two facts has made this system quite challenging to solve numerically.

Major breakthroughs in the solution of the model occurred during the 1960s, particularly in the one-dimensional steady-state case. A major factor was the research of H. K. Gummel, who, in two significant papers [54], [117], the second with D. L. Scharfetter on the transient model, introduced two decisive ideas, which have permeated this subject in the intervening years:

I System Decoupling (Nonlinear Gauss-Seidel Iteration);
II Exponential Fitting for the Continuity Equations.

The system decoupling map is a compact continuous map whose fixed points define solutions of the drift-diffusion model. The study of this map, or family of maps, as carried out in Chap. 4, comprises one of the fundamental objectives of this monograph. The Scharfetter–Gummel discretization has been employed widely in the numerical simulations of this model. Two-dimensional versions depend upon area element methods and subsequent reductions to one-dimensional discretizations along boundaries. For the reader who has not seen these methods, we describe the one-dimensional version in §2.4 below. In one dimension, the method proceeds by approximating the flux (current) by a piecewise constant function, where the breaks correspond to the nodal points. The concentration values at the nodal points are determined from a well defined linear system, the definition of which depends upon the numerical method selected. The solution of the linear system is equivalent to determining the constants in an exponential fitting method, arising from the exact solution of the locally constant flux equations.

The analysis of the multi-dimensional boundary-value problem in steady-state is complicated by the presence of mixed boundary conditions. On the contact portions of the device, including the three terminals and a grounded contact, Dirichlet boundary conditions are imposed. On the complement of this set within the device boundary, insulation conditions, or homogeneous Neumann boundary conditions, are imposed. The reader may recall that this is a situation in which gradient singularities may occur at the boundary transition points, say, in two dimensions. For example, Laplace's equation on the first quadrant quarter plane, where the segment $y = 0,\ 0 < x < 1$ is insulated, the semi-infinite segment $y = 0,\ x > 1$ is maintained at solution value unity, and the remaining boundary values are homogeneous Dirichlet values, leads to a gradient square-root singularity at $x = 1, y = 0$. This is easily seen by use of conformal mapping techniques, wherein the quarter plane is mapped onto a semi-infinite strip, via the transformation, $z = \sin w$. The solution is given by,

$$u(x, y) = \frac{2}{\pi} \arcsin \frac{1}{2} \left[\sqrt{(x+1)^2 + y^2} - \sqrt{(x-1)^2 + y^2} \right].$$

Separation of variables' techniques, applied to bounded region formulations, say, on the semi-disk, lead to similar results. This general topic has received considerable attention in the literature (see [7, 81, 137] for asymptotic developments and [106] for gradient integrability) in reference to linear elliptic equations. The results of [106] cover the case of three-dimensional devices, but are unduly pessimistic in allowing worst case scenarios.Thus, for example, included are singularities induced by line charges, a situation which does not correspond to the boundary conditions of our models. A careful two-dimensional study of asymptotic behavior for the device model has been carried out in [42], which includes the case of a singularity at a corner formed by the oxide region of a MOS-FET device.

The first mathematical treatment of the drift-diffusion model is due to Mock [103], who studied the steady-state case. Mock made essential use of the decoupling, as originally introduced by Gummel, in which the potential becomes the unknown mapping variable, and the quasi-Fermi levels are computed as a fractional step. In this formulation, the potential and the quasi-Fermi levels are the dependent variables of the model. The concentrations are related to the quasi-Fermi levels and the potential by exponential relations. A generalization of this work appeared in [119]. Neither of these authors, however, adequately modeled the carrier drift velocity , defined as the product, $\mu\mathbf{E}$, of the carrier mobility and electric field, respectively. Because of the phenomenon of saturation, it is required that the drift velocity be a bounded function of $\mathbf{E}$. However, since $|\mathbf{E}|$ can be unbounded near boundary transition points, it is therefore necessary that the decrease of μ compensate. This compensation induces a breakdown of uniform ellipticity when the Einstein relations are employed. The latter relate diffusion to mobility and temperature. The arguments required for this fairly delicate case were first introduced in [67], and are discussed in Chap. 4 below. An earlier, less technical version, appeared in [10]. In all cases, the Gummel fixed point map plays a central role. This name has come to signify a loose collection of maps, all defined by an appropriate decoupling procedure, in which some subset of the variables becomes the collection of mapping variables. This treatment will emphasize the quasi-Fermi levels as the mapping variables.

During the 1970s and early 1980s, Newton outer iteration procedures began to complement successive approximation based upon iteration of the Gummel map. Major computational breakthroughs occurred when effective inner iteration was designed in tandem with the outer iteration. Noteworthy was the paper [13], in which damped Newton outer iteration, a form of continuation, was combined with Gauss-Seidel and other forms of inner iteration, thus replacing sparse direct linear algebra methods with iterative methods. This paper also presents a two-dimensional Scharfetter–Gummel discretization. The difficult implementation problem in three dimensions appears to be the geometric partitioning problem, carried out so that the linear systems for the Poisson equation are of M-matrix type. Nonetheless, in recent years, three-dimensional simulators have been announced. These include the FIELDAY, TOPMOST, and CADDETH simulation programs, respectively (see [22, 121, 133]). In addition to these industrial device simulators, there is the two-dimensional simulator, PISCES II, released in 1985 by Stanford University, and the three-dimensional simulators, SIMUL_ISE and MINIMOS, developed at the ETH, Zurich, and the Technical University of Vienna, respectively. Robert Dutton, Wolfgang Fichtner, and Siegfried Selberherr have been the leaders in the development of the university simulators just cited.

Two books have appeared in the mathematical literature on the drift-diffusion model: [100, 104]. The latter also discusses the transient problem. Three recent conference proceedings related to device and/or circuit modeling are currently available: [11, 12, 57]. Although the transient model is not discussed in the current book, the reader may consult [30] and [69] for development. A book which incorporates models of quantum transport is [101].

We shall close this section by making some comments on the drift-diffusion, hydrodynamic, Boltzmann hierarchy and connections to related models. More details are furnished in the following chapter. The unknown in the Boltzmann transport equation is a density function, $f(\mathbf{x}, \mathbf{u})$ defined on physical space and group velocity space. Simplifications of Boltzmann's equation are usually sought. These can be obtained by integration over velocity space, after which only a density in physical space remains. Obviously, a major amount of information is lost in this reduction process. In order to make up for this loss, more than one "moment" of Boltzmann's equation can be taken. Simple integration of f over velocity space yields the zeroth moment, while integration of the product of f by the group velocity yields the first moment. Integration of f times the energy yields a second moment, and so on. In the averaging process over velocity space, the response of the electron density and the electron energy density to electric fields is summarized in mobility functions. These are discussed at the end of this chapter. It should be mentioned, however, that certain semiconductor devices, such as GaAs devices, probably are not accurately modeled by scalar mobility functions (cf. [58]) within the context of the current continuity equations alone.

The collisionless Boltzmann transport equation is called the Vlasov equation. This equation is derived from a so-called master equation of high dimension, the Liouville equation, which is equivalent to the Hamiltonian formulation in mechanics. Long range Coulomb forces are incorporated into the Vlasov model via the Poisson equation, but short range forces lead to the collision operator in the Boltzmann transport equation. This equation is called semi-classical if the collision effects are compatible with quantum principles. There are also the quantum Liouville equation and the quantum Boltzmann equation, and transforms of these equations involving the Wigner function, a type of distribution function, which need not be positive, however. These ideas are described in [101].

2.3 Scaling and Junction Width Estimation

We shall introduce a special version of the drift-diffusion system in this section for the purpose of explaining, through the choice of well known scalings, the highly convective character of the continuity equations, and the possible singular perturbation aspect of the potential equation. This includes the determination of the junction width in an accompanying example as proportional to the square root of the singular perturbation parameter.

2.3.1 System and Scalings

The simplifications employed in this subsection are as follows:

a) Carrier recombination is suppressed.
b) The dielectric is assumed constant over the device.
c) The mobility and diffusion coefficients are assumed constant.

The unscaled Poisson equation is given by

$$-\nabla\cdot(\epsilon\nabla\phi) = e(k_1 - n + p), \tag{2.1}$$

where the dielectric constant ϵ is set equal to the product of the dielectric constant ϵ_0 for the vacuum times the relative dielectric constant of the semiconductor that is modeled, k_1 is the net impurity concentration (doping), and e is the magnitude of the electronic charge. The dependent variables are the electrostatic potential, ϕ, and the electron and hole concentrations, n and p. This equation is the Poisson electrostatic equation, and the right hand side represents the net charge density. Typical values for the associated constants are:

$$\epsilon_0 = 8.85\times 10^{-14}\ \mathrm{F/cm},$$
$$e = 1.6\times 10^{-19}\ \mathrm{C},$$

for the constants of nature, and for the material constants, one has

$$\epsilon_{\mathrm{Si}} = 11.7,$$
$$\|k_1\|_{\max} = 10^{17} - 10^{20}/\mathrm{cm}^3\ .$$

The relative dielectric constants of other well known semiconductors are of the order of silicon.

The (steady-state) continuity equations are given by

$$\nabla\cdot\mathbf{J}_n = 0, \tag{2.2}$$
$$\nabla\cdot\mathbf{J}_p = 0. \tag{2.3}$$

Here the electron and hole current densities are given by the constitutive relations, involving a diffusion component and a drift component,

$$\mathbf{J}_n = e(D_n\nabla n - \mu_n n\nabla\phi), \tag{2.4}$$
$$\mathbf{J}_p = -e(D_p\nabla p + \mu_p p\nabla\phi). \tag{2.5}$$

The quantities μ_n and μ_p are the electron and hole mobility coefficients and D_n and D_p are the diffusion coefficients The Einstein relations, linking mobility and diffusion, are typically assumed in steady-state modeling:

$$D_n = (kT_0/e)\mu_n,$$
$$D_p = (kT_0/e)\mu_p,$$

where k is Boltzmann's constant and T_0 is the lattice temperature. The expression $U_T \equiv kT_0/e$ is called the thermal voltage; its value at $T_0 = 300\mathrm{K}$ is:

$$U_T = .0259\mathrm{V}.$$

Low field values of the mobilities for silicon are given by (at 300 K),

$$\begin{aligned} \mu_n &= 1350\mathrm{cm}^2/\mathrm{V}/\mathrm{sec}, \\ \mu_p &= 480\mathrm{cm}^2/\mathrm{V}/\mathrm{sec}. \end{aligned}$$

The mobility values are given for completeness; they are constant multipliers, eventually divided out, in the simplified continuity equations.

We are now ready to introduce two closely related scalings for the system (2.1–2.3). We shall refer to these as the De Mari and the unit scalings.

1. De Mari Scaling
   - Potentials are scaled by U_T: $u = \phi/U_T$.
   - All concentrations are scaled by the intrinsic (undoped) concentration, n_i, which for silicon has the value, $1.4 \times 10^{10}/\mathrm{cm}^3$.
   - Length is scaled by a characteristic length:

$$l = \sqrt{\epsilon U_T/(en_i)}.$$

2. Unit Scaling
   - Potentials are scaled by U_T.
   - All concentrations are scaled by $\|k_1\|_{\max}$.
   - Length is scaled by micron dimensions.

The new system can be written, in the case of constant mobilities and under the assumption that the Einstein relations are valid:

$$-\lambda^2\nabla^2 u + n - p = k_1, \tag{2.6}$$

$$-\nabla\cdot(\nabla n - n\nabla u) = 0, \tag{2.7}$$

$$-\nabla\cdot(\nabla p + p\nabla u) = 0. \tag{2.8}$$

In the case of each of the scalings,

$$\lambda^2 = \epsilon U_T/(l^2 eS),$$

where S is the appropriate concentration scale. In the case of the De Mari scaling, $\lambda^2 = 1$, whereas in the case of the unit scaling, $\lambda^2 \approx 10^{-1} - 10^{-7}$. When measured in De Mari units, the doping k_1 is of the order of $10^7 - 10^{10}$, and the electron and hole concentrations would be expected to have these maximal values also. In the unit scaling, so-called because the doping has maximal value 1 in these units, the potential equation (2.6) can be singularly perturbed, and all concentrations are expected to be of order comparable to unity.

One final remark about scalings is in order. It is not necessary to use any scaling at all, even in micron and submicron devices. What is essential is the proper choice of units in this case. One may view this procedure as self-induced scale, via unit selection. In [37], the following choices were made:

1. The length unit is the micron and the time unit is the picosecond.
2. Energy, charge, and capacitance units are 10^{-18} of the values of the Joule, Coulomb, and Faraday, respectively.
3. Potential is expressed in volts and temperature in degrees Kelvin.
4. Mass units are 10^{-30} of a kilogram.

Concentrations are thus reduced by a factor of a trillion, due to the choice of volume units. In [37], simulations were effectively carried out for the $n^+ - n - n^+$ diode, with maximum doping levels of $5 \times 10^{17}/\text{cm}^3$. In the units employed, such concentrations were expressed as 500 000 units.

2.3.2 Example and Heuristic Analysis

We shall present a very simple example to indicate the boundary layer possible near a p/n junction. Since such layers are associated with steep gradients, and since the electric field enters the formulations (2.7, 2.8), via the convective or drift terms, this has the simultaneous effect of demonstrating that the continuity equations can be highly convective. The calculation here is an adaptation of that in [129, pp.142–147]. Consider then a one-dimensional p/n metallurgical junction positioned at $x = 0$. The junction is situated so that the p-region occupies a bounded part of $x < 0$ and the n-region occupies a bounded part of $x > 0$. The junction is assumed to be a so-called step junction:

$$k_1 = \begin{cases} -N_a, \ x < 0, \\ N_d, \ x > 0. \end{cases}$$

The calculations are based upon the following *idealized* assumptions:

1. The device is in equilibrium, i.e., the electron and hole current densities are zero. In particular, the (far) boundary conditions are determined as follows.
   – There are no external voltage biases, and the electron and hole concentrations are specified in the usual way by thermal equilibrium and charge balance, respectively:

$$np = (n_i/S)^2 \quad =: \quad c^2, \tag{2.9}$$

$$n - p - k_1 \quad = \quad 0. \tag{2.10}$$

2. There is a carrier depletion region surrounding the junction, written as $[-x_{p0}, x_{n0}]$, in which moment conditions hold related to the size of $n - p$:

$$\int_{-x_{p0}}^{x_{n0}} (n(x) - p(x))\, dx \approx 0, \tag{2.11}$$

$$\int_{-x_{p0}}^{x_{n0}} x(n(x) - p(x))\, dx \approx 0. \tag{2.12}$$

3. Adjacent to the depletion region, there is a neutral region, characterized by:
   – A zero electric field.

Physically, the depletion region occurs when electrons diffuse from the n-region to the p-region, and then some are swept by the field from p to n (conversely for holes). It is immediate from (2.11) that the charge on the p side of the depletion region balances that on the n side. This observation is used in (2.18) below. The boundary conditions, (2.9, 2.10), give the familiar thermal equilibrium relations for n and p:

$$n = \frac{1}{2}\left(k_1 + \sqrt{k_1^2 + 4c^2}\right), \tag{2.13}$$

$$p = \frac{1}{2}\left(-k_1 + \sqrt{k_1^2 + 4c^2}\right). \tag{2.14}$$

These relations are valid at the endpoints of the device, but also, because of the neutral region and equilibrium assumptions, in the entirety of the neutral region. Indeed, from the equilibrium assumption, we have the explicit solution relations, which also determine the boundary values of u,

$$n = c\exp(u), \quad p = c\exp(-u), \tag{2.15}$$

which, together with the constant values of the potential in the two components of the neutral region, yields the assertion. We define the contact potential by $u_n - u_p := u_0$. Here, u_n and u_p are the values of the potential at x_{n0} and $-x_{p0}$, respectively. Use of (2.13–2.15), together with a simple approximation based on the size of N_a and N_d, yields the expression,

$$u_0 = \log N_a N_d + 2\log(S/n_i). \tag{2.16}$$

Integration of the equation, $E = -du/dx$, across the depletion region gives the relation, for $W = x_{p_0} + x_{n_0}$,

$$u_0 = (1/(2\lambda^2))N_d x_{n_0} W. \tag{2.17}$$

This formula, obtained by integration by parts applied to the integrated version of (2.6), makes use of (2.12). The unknown quantities are x_{n_0} and W. The two equations specifying these quantities are (2.17) and the charge balance relation:

$$N_a(W - x_{n_0}) = N_d x_{n_0}. \tag{2.18}$$

Note that u_0 is specified by (2.16). Solution of these relations gives, finally, the formula for the width of the depletion region:

$$W = \lambda\sqrt{2u_0(N_a + N_d)/(N_a N_d)}. \tag{2.19}$$

Suppose now, following [129], we set $N_d = 7.1 \times 10^5$ and $N_a = 2.9 \times 10^8$ intrinsic concentration units. Then $W = 9.6 \times 10^{-3}$ length units, which in this case are 34.7 microns. If a piecewise linear approximation for E is employed, yielding the relation, $u_0 = -\frac{1}{2}E_0 W$, then $E_0 = -6866$. The values in the unit scaling are $W = .33$ and $E_0 = -199.7$. In the piecewise linear approximation, E is an inverted sawtooth function on the depletion region, and we see that the convection coefficient varies between 0 and E_0 on this region of width W.

2.4 The Scharfetter–Gummel Discretization

In this section, we illustrate the class of discretizations of convection-diffusion equations which is known by the name of Scharfetter–Gummel discretizations. As shown below in §2.4.2, the class is described in one dimension by the property:

– The associated flux of the numerical approximation is piecewise constant.

Furthermore, the procedures for selecting the nodal values of the approximation differentiate methods within the class. We begin with a subsection in which variational principles are discussed in N dimensions for convection-diffusion boundary-value problems. The reader should understand that we do not intend in this section to make general mathematical statements in the form of theorems.

2.4.1 Variational Calculus

We shall restrict attention to single equations of the form

$$-\nabla\cdot \mathbf{J} = F \text{ in } G, \tag{2.20}$$

$$n = \bar{n} \text{ on } \partial G, \tag{2.21}$$

where $\mathbf{J}$ is specified by

$$\mathbf{J} = \nabla n - n\nabla u. \tag{2.22}$$

Here, u, F, and $\bar{n}$ are given functions. We shall consider those cases where $\bar{n}$ is strictly positive or identically zero. Under suitable regularity assumptions on u and $\bar{n}$, unique weak solutions of (2.20–2.22) can be shown to exist, via a minimization principle, as demonstrated in Chap. 4 in a more general context. This principle can then serve as the basis for a Ritz-Galerkin approximation theory if desired. It is instructive to apply the superposition principle, so that the above Dirichlet problem is decomposed into the two subproblems, first, with homogeneous right hand side ($F \equiv 0$) and second, with homogeneous boundary data ($\bar{n} \equiv 0$), respectively. The first of these is analyzed by a classical exponential change of variable:

$$n = \exp(u - v). \tag{2.23}$$

The function v here is purely generic, but in the application to the drift-diffusion model it is known as a quasi-Fermi level. The corresponding quasi-Fermi level for the holes is obtained in the sequel through the introduction of a function w and the change of variables,

$$p = \exp(w - u). \tag{2.24}$$

In the exponential variable, $\mathbf{J}$ is represented by

$$\mathbf{J} = -\exp(u - v)\nabla v. \tag{2.25}$$

When the substitution $V = e^{-v}$ is made in (2.25), one obtains a relation linear in V, and the solution of the first subproblem is characterized as the solution of the quadratic extremal problem,

$$\min_{\mathcal{F}} \int_G |\mathbf{J}|^2 e^{-u}, \tag{2.26}$$

where $\mathcal{F}$ is an affine space:

$$\mathcal{F} = \{V = \bar{V} + \nu : \nu \in H_0^1\},$$

and $\bar{V}$ is determined by $\bar{n}$ via $\bar{n} = \bar{V}e^u$. Of course, the solution is directly exhibited in terms of the new variable, V.

The solution of the second subproblem need not entail the change of variable. Indeed, the operator L defined on C_0^∞ by

$$Ln = -\nabla \cdot \mathbf{J}, \tag{2.27}$$

satisfies the following important property:

L is symmetric in the (real) weighted $L_{2,\omega}$ inner product defined by $\omega = e^{-u}$ and

$$(f, g)_{L_{2,\omega}} = \int_G fge^{-u}, \tag{2.28}$$

i.e., the relation,

$$(Lf, g)_{L_{2,\omega}} = (f, Lg)_{L_{2,\omega}}, \tag{2.29}$$

holds on the domain of L.

This follows by direct computation. It can be shown that the closure of L is a nonnegative definite, self-adjoint operator in $L_{2,\omega}$, with domain $H^2 \cap H_0^1$. This follows directly by integrating the left hand side of (2.29) by parts when $g = f$. The nonnegative definite property can be strengthened to positive definiteness if there is no zero eigenvalue. This translates into the condition,

$$\mathbf{J} \equiv 0,\ n \in H_0^1 \Rightarrow n \equiv 0. \tag{2.30}$$

Under this condition, it is known from the Riesz/Fredholm/Schauder theory (cf. [6, Theorem 2.3, p. 181]) that the unique solution of the second subproblem is characterized by projection. In this case, projection is simply the gradient form of the minimization problem,

$$\min_{n \in H_0^1} \left\{ \int_G |\mathbf{J}|^2 e^{-u} - 2(F, n)_{L_{2,\omega}} \right\}. \tag{2.31}$$

The verification of these statements is carried out by standard arguments from variational theory and the theory of weak solutions (cf. [48] and [102]), or by specialization of the general arguments of Chap. 4.

2.4.2 Piecewise Constant Flux

We consider the one-dimensional version of (2.20–2.22), and require that the scheme be exact for a piecewise constant flux J whose discontinuities coincide with selected grid points; note that our interpretation of solution will be that of a weak solution throughout. For simplicity of exposition only we shall assume that $\bar{n} \equiv 0$. If the exactness requirement is interpreted in terms of approximation theory, we are seeking an approximation of the form

$$n_h = \sum_i \alpha_i M_i, \tag{2.32}$$

where $\{\alpha_i\}$ is a set of nodal values determined by a specified numerical method, and where $\{M_i\}$ is a nodal basis of local support functions defined by L. More precisely, given a grid of the interval $\bar{G} = [0,1]$, of the form $x_i = ih$, $i = 0, \cdots d$, for $dh = 1$, M_i is associated with the ith grid point, $i \neq 0$ and $i \neq d$, and is specified by the following requirements:

1. $M_i(x_i) = 1$, support $M_i = [x_{i-1}, x_{i+1}]$.
2. M_i is continuous.
3. On each subinterval determined by the grid, M_i is in the null space of L.

We shall refer to discretizations of the form (2.32) as being of the class of Scharfetter–Gummel discretizations. We shall comment later on the critical issue of nodal value determination. The functions M_i are generalizations of the chapeau functions, and are examples of the generalized B-splines introduced in [63] and studied extensively in [79]. The approximation given by (2.32) is an example of a generalized spline, introduced in [51]. It is quite easy to give explicit formulas for the B-spline functions, M_i. For example, when $i = 1$,

$$M_1(x) = \begin{cases} \exp[u(x) - u(x_1)] \int_{x_0}^{x} e^{-u(s)}\, ds / \int_{x_0}^{x_1} e^{-u(s)}\, ds, & x_0 \le x \le x_1, \\ \exp[u(x) - u(x_1)] \int_{x}^{x_2} e^{-u(s)}\, ds / \int_{x_1}^{x_2} e^{-u(s)}\, ds, & x_1 \le x \le x_2. \end{cases}$$

The general formula is obtained via the identifications $0 \to i-1$, $1 \to i$, and $2 \to i+1$. In all cases, the definition is completed by the support requirement in the second part of item one above.

It is also quite straightforward to compute the piecewise constant flux J. Note that, on the intervals (x_0, x_1) and (x_{d-1}, x_d), the flux is determined by M_1 and M_{d-1}, respectively, while on all other subintervals, (x_i, x_{i+1}), the flux has components from both M_i and M_{i+1}. The total flux can then be assembled from the following result. Denoting by $J^-_{M_i}$ the flux due to M_i on (x_{i-1}, x_i), and by $J^+_{M_i}$ the flux due to M_i on (x_i, x_{i+1}), we have

$$\begin{aligned} J^-_{Mi} &= e^{-u(x_i)} \Big/ \int_{x_{i-1}}^{x_i} e^{-u(s)}\, ds, \\ J^+_{M_i} &= -e^{-u(x_i)} \Big/ \int_{x_i}^{x_{i+1}} e^{-u(s)}\, ds. \end{aligned}$$

In order to evaluate the integrals appearing in the flux representations, it has been common to employ the piecewise linear interpolant of u. When this is done, direct flux evaluation gives the following representation for the assembled flux on (x_i, x_{i+1}), with $\alpha_i = n_i$:

$$J = (1/h)[B(\Delta u)n_{i+1} - B(-\Delta u)n_i)], \tag{2.33}$$

where we have adopted the conventions $\Delta u = u(x_{i+1}) - u(x_i)$ and $B(z) = z/[\exp(z) - 1]$. The latter function, known as the Bernoulli function, must be computed for small values of $|z|$ by series methods involving the Bernoulli numbers. As noted in [100], an exponential fitting method of this type resolves the currents adequately, even if the mesh allows for substantial variation in the function u.

Although the form given in (2.33) is well-known to workers in computational electronics, it is not widely understood that the form holds, irrespective of the numerical method used to characterize the nodal values. The original method of Scharfetter and Gummel was to define these values by the box method, or, in current parlance, a finite volume method. Such a method is tailored for divergence equations, and yields a nonsymmetric tridiagonal matrix to "invert." In the sequel, this approach is what will be meant when we speak of the Scharfetter–Gummel method. In this case, the matrix entries are given by

$$\begin{aligned} d_i &= B(\Delta u_i) + B(-\Delta u_{i+1}), \\ c_i &= -B(\Delta u_{i+1}), \\ a_i &= -B(-\Delta u_{i+1}), \end{aligned}$$

where the diagonal subscripts for d_i range from 1 to $d-1$ and the superdiagonal and subdiagonal subscripts for c_i and a_i, respectively, range from 1 to $d-2$.

An approximation procedure which may well be superior to the classical procedure just cited, and which is also of the class introduced in this section, is a simple Ritz procedure, based upon the variational principle discussed above for the second subproblem. In this procedure, the quadratic form defined in (2.31) is minimized over the space of generalized splines of the form (2.32). This gives rise to the following computational advantages:

1. The matrix problem is positive definite, symmetric, and,
2. A generalized Aubin-Nitsche duality argument (cf. [128]) holds, so that the order of convergence in the L_2 norm is two.

The (maximal) order for the classical Scharfetter–Gummel method has not been demonstrated. This appears to be the first time this particular Ritz-Galerkin procedure has been proposed, although various forms of related Petrov-Galerkin methods have appeared in the literature. This is, indeed, a Petrov-Galerkin method in the usual sense, but is here shown to have much more tangible structure than typical such methods, in that it is identified as a Ritz method in a weighted inner product space. A specialized form of the trial space introduced here was considered in [34].

2.5 A Model for Drift-Diffusion

The drift-diffusion model may be obtained by taking zeroth order moments of the BTE and adjoining the Poisson equation. Thus, one obtains the system for M carriers with concentration n_i, carrier recombination R_i, current density $\mathbf{J}_i$, (signed) charge e_i, $i = 1, \cdots, M$, with the units of k_1 now reflecting the charge modulus:

$$\frac{e_i \partial n_i}{\partial t} + \nabla \cdot \mathbf{J}_i = -e_i R_i, \tag{2.34}$$

$$\mathbf{E} = -\nabla \phi, \tag{2.35}$$

$$\nabla \cdot (\epsilon \nabla \phi) = -\sum e_i n_i - k_1. \tag{2.36}$$

There still remains the issue of determining the constitutive current relations. Classical drift-diffusion theory gives, for $M = 2$, $n_1 = n$, and $n_2 = p$,

$$\mathbf{J}_n = eD_n \nabla n - e\mu_n(\mathbf{E}) n \nabla \phi, \tag{2.37}$$

$$\mathbf{J}_p = -eD_p \nabla p - e\mu_p(\mathbf{E}) p \nabla \phi. \tag{2.38}$$

Here e is positive. Notice that the drift terms,

$$-e\mu_n(\mathbf{E}) n \nabla \phi, \; -e\mu_p(\mathbf{E}) p \nabla \phi,$$

are proportional to the respective drift velocities,

$$\mu_n(\mathbf{E}), \; \mu_p(\mathbf{E})$$

By experimental observation, these velocities are bounded as functions of $|\mathbf{E}|$. In fact, the asymptotic limit is given in (2.43).

The introduction of exponential relations for n and p is also common (cf. §2.4), as is the use of the Einstein relations linking the mobility coefficients μ_n, μ_p, and the diffusion coefficients D_n, D_p. These relations are specified by

$$D_n = (kT/e)\mu_n, \tag{2.39}$$

$$D_p = (kT/e)\mu_p. \tag{2.40}$$

Thus, if the dimensionless potential $u = \frac{\phi}{kT/e}$ is introduced, and the Einstein relations are employed, then (2.37, 2.38) simplify considerably; moreover, if one employs the dimensionless quasi-Fermi levels v and w, so that $n = \exp(u - v)$ and $p = \exp(w - u)$, then

$$\mathbf{J}_n = -e\mu_n \exp(u - v) \nabla v, \;\; \mathbf{J}_p = -e\mu_p \exp(w - u) \nabla w. \tag{2.41}$$

It is also possible to derive the constitutive relations (2.37, 2.38) from the first order moment relations under the assumption that the momentum relaxation times tend to zero. The details are given in [116]. In fact, the constitutive relations include a heat flux term as well, which is suppressed at constant temperature. If it is not suppressed, one has what is called an energy transport model. In this derivation, one uses the definition of mobility in terms of relaxation time, to be presented in the next subsection. This discussion is amplified in the next chapter. Finally, it is common in the classical drift-diffusion model to take for the carrier recombination the Shockley-Read-Hall-Auger term R, for both the electron and hole equations. However, since the electron equation has a factor $-e$ (cf. (2.34)), the net result is that this term appears oppositely signed in the equations. The form of this function is made explicit in (4.70) in Chap. 4. Note that the factor, $\exp(w - v) - 1$, is critical for the analysis.

2.5.1 The Mobility Relations

The carrier mobility has the physical units of $\mathrm{cm}^2\mathrm{V}^{-1}\mathrm{sec}^{-1}$. In terms of the momentum relaxation time, it is characterized by

$$\mu = e\tau_p/m.$$

As was noted in §2.2, the definition of mobility, even for low order moment models such as the drift-diffusion model, can be enhanced by use of the higher order moment models. For example, we briefly distill here the arguments of [56] to derive the field dependent expression for the mobility given by

$$\mu(\mathbf{E}) = 2\mu_0 / \left[1 + \sqrt{1 + 4(\mu_0|\mathbf{E}|/v_d)^2}\right]. \tag{2.42}$$

Here, μ_0 and v_d are constant only with respect to the electric field. The latter quantity has the interpretation of saturation velocity:

$$\lim_{|\mathbf{E}|\to\infty} \mu(\mathbf{E})|\mathbf{E}| = v_d. \tag{2.43}$$

According to [56], the momentum relaxation time is expressed, by taking contributions up to third moments, in terms of the relation,

$$\tau_p^{-1} = \tau_{p0}^{-1}\left(1 + \eta\frac{W - W_0}{n}\right). \tag{2.44}$$

In this relation, the quantities τ_{p0} and η are independent of the electric field. The approach of [56] is to characterize (2.42) in terms of homogeneous bulk material with constant electric field. In this situation, the momentum and energy equations reduce to the following relations:

$$\mathbf{J} = e\mu n\mathbf{E}, \tag{2.45}$$

$$(W - W_0)/\tau_w = \mathbf{J}\cdot\mathbf{E}. \tag{2.46}$$

By beginning with the definition of μ, and employing (2.44– 2.46), we obtain a quadratic equation for μ. By taking the positive root, and redefining in terms of

$$\begin{aligned} \mu_0 &= e\tau_{p0}/m, \\ v_d &= 1/\sqrt{\eta\tau_w m}, \end{aligned}$$

we obtain (2.42). In certain places in this book, we shall make use of this or similar relations for the mobility coefficients.

2.5.2 Boundary Conditions and Current-Voltage Relations

A careful discussion of the various types of boundary conditions is given in [120] and [130]. These are strongly dependent on the type of contact used in the particular device; ohmic and Schottky contacts are typically employed. On the remainder of the boundary, it has been traditional to require zero flux for the electric field, and for the carrier current densities. This is logically equivalent to standard homogeneous Neumann boundary conditions for the dependent variables on this portion of the boundary. It should be emphasized, however, that these traditional choices are idealizations at best, since adequate consideration of effects external to the device is missing. For ohmic contacts, the relations (2.9) and (2.10) define the boundary conditions on the Dirichlet boundary Σ_D of the device. These give the explicit relations, (2.13) and (2.14). When the Fermi levels are employed, the boundary values are specified via (2.13) and (2.14), once the boundary values for the potential are clearly defined. It is customary to describe the boundary conditions for the potential via two terms, a built-in potential, u_{bi}, due to the doping, and an applied bias potential, which may depend on time and position. The built-in potential is defined as the equilibrium potential, corresponding to zero carrier currents. This was the case analyzed in §2.3.2 in one dimension for a single junction. Thus, if we use the equation for n defined by (2.15), and the boundary-value equation for n defined by (2.13), we obtain

$$u_{bi} = \ln\left(\frac{1}{2c}\left(k_1 + \sqrt{k_1^2 + 4c^2}\right)\right). \tag{2.47}$$

In MOS-FET devices with an oxide region, it is customary to consider the Poisson equation on the union of the substrate and buffer, with continuous electric displacement vector across the common boundary, and to consider the current continuity subsystem on the substrate alone, with a zero current flux for the carriers across the substrate/buffer boundary.

The current-voltage curves in a MOS-FET device are derived from plotting the line integral, $I = \int_C \mathbf{J} \cdot ds$, as a function of the applied source-drain bias, V. Here, C is a curve enclosing the drain in the two-dimensional device. The behavior of the current-voltage curve is fundamental in semiconductor device theory. It is a nonlinear version of Ohm's law for the response of the device. A striking property of quantum devices, separating them from standard devices, is that the current-voltage curve can develop so-called regions of negative differential resistance, where the current falls with increasing voltage, and can even develop hysteresis loops (bistable states). Resonant tunneling diodes exhibit this kind of behavior.

3. Moment Models: Microscopic to Macroscopic

3.1 The Hydrodynamic Model

In this chapter, we shall present alternative models to that of drift-diffusion. More precisely, the models developed here are refinements, for which drift-diffusion is a limiting case. The two which we select are the classical hydrodynamic model, and a class of energy transport models. A quantum transport variant of the hydrodynamic model is briefly discussed at the conclusion of the chapter. For the reader who wishes to pursue these topics in greater detail, a special issue of *VLSI DESIGN*, edited by the author, has appeared (vol. 3, no. 2, 1995).

We begin with the hydrodynamic model, which is discussed in the remainder of this section and in §3.3. The energy transport model is discussed in §3.4, and the quantum hydrodynamic model is discussed in §3.6.

3.1.1 Charge, Momentum and Energy Transport Equations

The physical perspective adopted here is that the electron gas behaves, with respect to its averaged properties, like a dilute gas, as studied in the kinetic theory of gases. The critical issue of collision mechanisms is resolved in terms of relaxation expressions. The most comprehensive mathematical survey of this material is included in the recent book [24]. The equations as presented are discussed in references [16, 23, 116]. The first and third references are specific to semiconductors, while the second is quite general, and omits some forcing terms. These equations are, respectively, derived as zeroth, first, and second order moments of the Boltzmann transport equation, with the latter written for an electron species moving in an electric field as

$$\frac{\partial f}{\partial t} + \mathbf{u}\cdot \nabla_{\mathbf{x}} f - \frac{e}{m}\mathbf{E}\cdot \nabla_{\mathbf{u}} f = C. \tag{3.1}$$

Here, $f = f(\mathbf{x}, \mathbf{u}, t)$ is the numerical distribution function of a carrier species, $\mathbf{x}$ is the position vector, $\mathbf{u}$ is the species' group velocity vector, $\mathbf{E} = \mathbf{E}(\mathbf{x}, \mathbf{t})$ is the electric field, e is the electron charge modulus, m is the effective electron mass, and C is the time rate of change of f due to collisions. Many different types of collisions can be identified, including electron-lattice and electron-ion collisions, and also inter-particle collisions of various types. A typical representation of one of these mechanisms, embodying the Pauli exclusion principle, is given by the nonlinear expression (cf. [101]),

$$\begin{aligned} C &= \int \{S(\mathbf{x},\mathbf{u}',\mathbf{u})f(\mathbf{x},\mathbf{u}',t)(1-f(\mathbf{x},\mathbf{u},t)) \\ &- S(\mathbf{x},\mathbf{u},\mathbf{u}')f(\mathbf{x},\mathbf{u},t)(1-f(\mathbf{x},\mathbf{u}',t))\}\, d\mathbf{u}', \end{aligned} \tag{3.2}$$

in terms of a scattering kernel S. In the Boltzmann transport equation above, it has been assumed that the traditional Lorentz force field does not have a component induced by an external magnetic field. The moment equations, which will be derived subsequently, are expressed in terms of certain dependent variables, where n is the electron concentration, $\mathbf{v}$ is the (translational) velocity, $\mathbf{p}$ is the momentum density, P is the symmetric pressure tensor, $\mathbf{q}$ is the the heat flux, e_I is the internal energy, and C_n, $\mathbf{C_p}$, and C_W represent moments of C, taken with respect to the functions,

$$\begin{aligned} g_0(\mathbf{u}) &\equiv 1, \\ g_1(\mathbf{u}) &= m\mathbf{u}, \\ g_2(\mathbf{u}) &= \frac{m}{2}|\mathbf{u}|^2. \end{aligned}$$

The moment equations are given by:

$$\frac{\partial n}{\partial t} + \nabla\cdot(n\mathbf{v}) = C_n, \tag{3.3}$$

$$\frac{\partial \mathbf{p}}{\partial t} + \mathbf{v}(\nabla\cdot\mathbf{p}) + (\mathbf{p}\cdot\nabla)\mathbf{v} = -en\mathbf{E} - \nabla\cdot\mathsf{P} + \mathbf{C_p}, \tag{3.4}$$

$$\begin{aligned} \frac{\partial}{\partial t}\left(\frac{mn}{2}|\mathbf{v}|^2 + mne_I\right) + \nabla\cdot\left(\mathbf{v}\left[\frac{mn}{2}|\mathbf{v}|^2 + mne_I\right]\right) + \nabla\cdot(\mathbf{v}\mathsf{P}) = \\ - en\mathbf{v}\cdot\mathbf{E} - \nabla\cdot\mathbf{q} + C_W. \end{aligned} \tag{3.5}$$

The Poisson equation for the electric potential must be adjoined. The reader should realize that each species contributes a corresponding moment subsystem, with appropriately signed charge. In the subsystem (3.3–3.5) above, it is understood that the tensor P acts like a matrix; left action by vectors denotes row vector multiplication, etc.

We begin the derivation with the definitions and assumptions standard in the kinetic theory of gases (see [60, 96]).

Definition 3.1.1 (macroscopic variables). *1. The concentration n is the integral of the Boltzmann distribution over group velocity space:*

$$n := \int f\, du.$$

2. *The velocity* $\mathbf{v}$ *is given by the average,*

$$\mathbf{v} := \frac{1}{n} \int \mathbf{u} f \, du.$$

3. *The momentum* $\mathbf{p}$ *is given by the product,*

$$\mathbf{p} := mn\mathbf{v}.$$

4. *The random velocity* $\mathbf{c}$ *is given by the difference,*

$$\mathbf{c} := \mathbf{u} - \mathbf{v}.$$

5. *The pressure tensor* P *is given by the second moments,*

$$P_{ij} := m \int c_i c_j f \, du.$$

6. *The internal energy density* e_I *is given by the random kinetic energy density,*

$$e_I := \frac{1}{2n} \int |\mathbf{c}|^2 f \, du.$$

   This function represents energy/unit mass/unit concentration.

7. *The heat flux* $\mathbf{q}$ *is given by the third moment,*

$$q_i := \frac{m}{2} \int c_i |\mathbf{c}|^2 f \, du.$$

The assumptions on f are now stated.

Assumption 3.1. The function f decreases sufficiently rapidly at infinity, or, more precisely, at the boundary of the Brillouin zone:

$$\lim_{|\mathbf{u}| \to \infty} g_i(\mathbf{u}) f(\mathbf{u}) = 0, \; i = 0, 1, 2.$$

Finally, we make certain observations about integral evaluations, which follow from the definitions and assumptions after integration by parts.

Property 3.1.1. (integral identities)

1. $\int c_i f \, du = 0.$
2. $\int \frac{\partial f}{\partial u_i} \, du = 0.$
3. $\int u_j \frac{\partial f}{\partial u_i} \, du = -\delta_{ij} n.$
4. $\int |\mathbf{u}|^2 \frac{\partial f}{\partial u_i} \, du = -2nv_i.$

The derivation of the conservation law equations (3.3–3.5) now continues as follows. We multiply the Boltzmann transport equation (3.1) by g_0, g_1, and g_2, respectively, and integrate over group velocity space in the variable $\mathbf{u}$; this is typically described by Brillouin zones, which here are assumed of infinite extent. The mass (equivalently, charge) conservation equation (3.3) is immediate from the definitions of $\mathbf{v}$ and C_n, and from the second integral identity of Property 3.1.1. In order to derive the momentum conservation equation (3.4), we begin with the identity,

$$\int u_i u_j f \, du = n v_i v_j + \int c_i c_j f \, du, \tag{3.6}$$

which follows from an expansion of the components of $\mathbf{u}$ in terms of those of $\mathbf{v}+\mathbf{c}$, and use of the first integral identity above. If (3.6) and the definitions of P and $\mathbf{C_p}$ are applied to the integrated product of (3.1) and g_1, one obtains (3.4), after an application of the third integral identity. The derivation of (3.5) begins with the identity,

$$(1/2) \int u_i |\mathbf{u}|^2 f \, du =$$

$$\frac{n v_i}{2} |\mathbf{v}|^2 + v_i n e_I + \sum_j v_j P_{ij}/m + (1/2) \int c_i |\mathbf{c}|^2 f \, du, \tag{3.7}$$

which follows again from the expansion of $\mathbf{u}$, from the definitions of e_I and P, and from the first integral identity. An application of (3.7), written in terms of the identity (3.6), and the definitions of $\mathbf{q}$ and C_W, to the integrated product of (3.1) and g_2, gives (3.5), after an application of the fourth integral identity. This completes the mass/momentum/energy system derivation. In addition to these transport equations, we have the Poisson equation for the electric field, where $k_1 :=$ doping and $\epsilon :=$ dielectric:

$$\mathbf{E} = -\nabla\phi, \tag{3.8}$$

$$\nabla \cdot (\epsilon \nabla \phi) = -\sum e_i n_i - k_1. \tag{3.9}$$

Here, we have used the convention that there are different species, each of concentration n_i and (signed) charge e_i. The entire system consists of equations (3.3–3.5), repeated according to species, and (3.8, 3.9). For convenience, the units of k_1 reflect the charge modulus.

3.1.2 Moment Closure and Relaxation Relations

The system derived in the preceding subsection has fifteen dependent variables in the case of one species in physical space, determined by ϕ, n, $\mathbf{v}$, P, e_I, and $\mathbf{q}$. By moment closure is meant the selection of compatible relations among these variables, so that the number of equations is equal in number to the remaining primitive variables selected. One way of proceeding is to introduce a new tensor variable T, the effective carrier temperature, defined by the ideal gas law relationship,

$$P_{ij} = nkT_{ij},$$

where k is Boltzmann's constant, and a scalar variable W, the total carrier energy. A program of reduction to the basic variables n, v, W, and ϕ can be implemented by the following assumptions:

Assumption 3.2 (closure assumptions). 1. The pressure is isotropic, with diagonal entries P_s and off-diagonal entries zero, for a suitable scalar function P_s. By previous relations, P_s is related to e_I via $mne_I = \frac{3}{2}P_s$. In particular, this is a model without viscosity, defined as a perturbation of the pressure tensor. Notice that P_s has the interpretation of a *specific* pressure, i.e. , pressure per unit mass.

2. It follows from the previous assumption that temperature may be represented by a scalar quantity T, and that the internal energy is represented in terms of T by

$$me_I = \frac{3}{2}kT.$$

3. The energy density (per unit concentration) w is given by combining internal energy and parabolic energy bands:

$$w = me_I + \frac{1}{2}m|\mathbf{v}|^2,$$

and the total energy (per unit volume) W is the product, $W = nw$.

4. The heat flux is obtained by a differential expression involving the temperature. A popular such choice is the Fourier law,

$$\mathbf{q} = -\kappa\nabla T.$$

Here, κ is the thermal conductivity governed by the Wiedemann-Franz law(cf. [15] and (3.71)). Other choices include a convective term in the form of $\mathbf{q}$ (see [126]).

In the case of M species, the closure relations determine $(N+2)M+1$ variables in N spatial dimensions. One can rewrite the system (3.3–3.5) with the closure assumptions incorporated. We have the following equations for the hydrodynamic (HD) model:

$$\frac{\partial n}{\partial t} + \nabla\cdot(n\mathbf{v}) = C_n, \tag{3.10}$$

$$\frac{\partial \mathbf{p}}{\partial t} + \mathbf{v}(\nabla\cdot\mathbf{p}) + (\mathbf{p}\cdot\nabla)\mathbf{v} = -en\mathbf{E} - \nabla(nkT) + \mathbf{C_p}, \tag{3.11}$$

$$\frac{\partial}{\partial t}W + \nabla\cdot(\mathbf{v}\,(W + nkT)) = -en\mathbf{v}\cdot\mathbf{E} + \nabla\cdot(\kappa\nabla T) + C_W. \tag{3.12}$$

The final step deals with the replacement of the collision moments. Motivated by the approachs of [9, 56, 109, 116], we define the recombination rate R and the momentum and energy relaxation times, τ_p and τ_w, respectively, in terms of averaged collision moments as follows.

Assumption 3.3 (macroscopic relaxation approximation). 1. The particle recombination rate R is given via

$$R := -C_n := -\int C\,du.$$

2. The momentum relaxation time τ_p is given via

$$\frac{\mathbf{p}}{\tau_p} := -\int m\mathbf{u}C\,du := -\mathbf{C_p}.$$

3. The energy relaxation time τ_w is given via

$$-\frac{W - W_0}{\tau_w} := \frac{m}{2}\int |\mathbf{u}|^2 C\,du := C_W.$$

Here, W_0 denotes the rest energy, $\frac{3}{2}nkT_0$, where T_0 is the lattice temperature. Also, for single carrier applications, $C_n = 0$. Note that the relaxation time reciprocals appearing here have the units of frequencies, and suggest time scales over which collisions are likely to occur in influencing momentum or energy. The forms for the relaxation times used in [9] on the basis of higher order moments, and retained by subsequent authors, are:

$$\begin{aligned}\tau_p &= c_p/T,\\ \tau_w &= c_w\frac{T}{T+T_0} + \frac{1}{2}\tau_p.\end{aligned}$$

Here, c_p and c_w are physical constants. A comprehensive discussion of these issues is outside the scope of the current exposition. We recall, however, the representations of the relaxation times given in [56], used in our mobility derivation of the previous chapter. Note that the full model requires adjoining the Poisson equation, (3.9). It is of interest that numerical simulations can be based upon this model; for the $n^+ - n - n^+$ diode see [37] and [47]. The former constitutes the first use of shock capturing methods for the HD model of semiconductors.

The closure assumptions discussed above have two noteworthy characteristics:

- The ideal gas law is induced for the carriers by the definition of T. This may be interpreted as a balance of random kinetic energy and thermal energy.
- Average collision mechanisms principally describe electron-lattice and electron-ion interactions, and, in the momentum system, have the form of Stokes' law for damping (friction).

These issues, and the attendant suppression of viscosity are discussed at greater length in [35] and in [60]. In the former work, it is mentioned that numerical simulations reveal the isotropic and small viscosity assumptions made for the pressure tensor to be valid for present device simulation. The high frequency modes for small disturbance traveling wave solutions are computed analytically in [35], and reveal strong damping dependence on heat conduction and relaxation.

3.2 Calibration with the Mechanics of Charged Fluids

In this section, we shall discuss an independent derivation of the equations (3.10–3.12). This is the traditional fluid mechanics (macroscopic) derivation for charged fluids, i.e., under the influence of electrostatics. Importantly, this method confirms the moment method of the previous section.

We begin with a general conservation principle, which provides the basis of the derivation of both (3.10) and (3.12). Suppose a scalar quantity ρ is associated with a fluid, occupying a spatial domain G, where ρ is a volume density. Suppose that the (outward) flux of the fluid quantity across the boundary of a fixed spatial region $\mathcal{B} \subset G$ is determined as $\mathbf{J}_\rho$. By means of an application of the classical divergence theorem (cf. [40]), we then have the volume balance,

$$\int_{\mathcal{B}} -\frac{\partial \rho}{\partial t}\, dV = \int_{\mathcal{B}} \{\nabla \cdot \mathbf{J}_\rho + g\}\, dV, \tag{3.13}$$

where g is negatively signed for a source and positively signed for a sink. Since the region $\mathcal{B}$ is arbitrary, the integrands must agree in (3.13). We thus have the *conservation equation*,

$$\frac{\partial \rho}{\partial t} + \nabla \cdot \mathbf{J}_\rho + g = 0. \tag{3.14}$$

This equation holds in the entire region G of the fluid.

3.2.1 Conservation of Mass and Energy

If we select the electron concentration as the first scalar quantity, i.e., $\rho = n$, the corresponding electron flux is just

$$\mathbf{J}_n = n\mathbf{v},$$

where $\mathbf{v}$ is the velocity. This gives the equation (3.10), via (3.14), if we set $g \equiv 0$.

The energy conservation equation (3.12) is similar. As in the previous section, we define $W = mn\left(\frac{1}{2}|\mathbf{v}|^2 + e_I\right)$, as the "conserved" scalar quantity. The energy flux, $\mathbf{J}_w$, has two principal components:

– Adiabatic component, due to mass transfer only:

$$\mathbf{J}_a = \mathbf{v}\left(\frac{1}{2}n|\mathbf{v}|^2 + ne_I + P\right). \tag{3.15}$$

Here, P is the mechanical specific pressure. In isentropic processes, $\mathbf{J}_a$ is the only component of the energy flux. The component is retained, even in nonadiabatic processes, because the formal transformation effected by the adiabatic process identifies it as due to mechanical effects, which persist even when dissipation is present. More precisely, the derivation of the adiabatic flux term is given in [93], by means of the adiabatic version of the second law of thermodynamics,

$$0 = TdS = de_I + PdV.$$

– Heat dissipation due to temperature equilibration:

$$\mathbf{J}_d = -\kappa\nabla T, \quad \text{Fourier law.} \tag{3.16}$$

The final step is the determination of the source/sink term. In this model, it consists of two parts, viz., the damping term (sink), and the electrical (Joule heating) term,

$$g_J = v\cdot(-en\mathbf{E}) = v\cdot\mathbf{J}_{\text{conv}}, \tag{3.17}$$

where $\mathbf{E}$ is the electric field, and $\mathbf{J}_{\text{conv}}$ is the so-called convection current. It is frequently designated by $\mathbf{J}$. Note that strict conservation of energy holds if the flux term $\mathbf{J}_d$ and the Joule heating term vanish. In this case the system entropy does not change. This is the classical isentropic case, and also assumes the absence of friction.

As is shown in [50, p. 55], the Joule heating arises from the electrical energy term,

$$W_{\text{elect}} = \frac{1}{2}\epsilon|\mathbf{E}|^2,$$

via the application of Maxwell's equations to

$$\frac{\partial}{\partial t}\int_B W_{\text{elect}}\, dV,$$

when an energy balance is calculated. The result of this calculation is:

$$\frac{\partial}{\partial t}\int_B W_{\text{elect}}\, dV = \int_B \mathbf{E}\cdot\mathbf{J}_{\text{conv}}\, dV,$$

leading directly to (3.17). Here, since we are considering electrostatics only, we neglect a surface rate of exchange of electromagnetic energy which is also present in the above equation. Moreover, an electromagnetic term is accordingly missing from the energy as well. The derivation given in [50] shows that, when relativistic and electromagnetic effects, due to moving frames and magnetic fields, are accounted for, then the Joule heating is modified as follows:

I $\mathbf{E}$ is replaced by the Lorentz force, $\mathbf{E} + \mathbf{v} \times \mathbf{B}$; and,
II $\mathbf{J}_{\text{conv}}$ is replaced by the conductive component of the current, $\mathbf{J} - \mathbf{J}_{\text{conv}}$.

The friction term is assumed to be proportional to the energy excess over the rest energy, and the proportionality factor, $\frac{1}{\tau_w}$, induces an energy dissipation rate. Finally, the assembled flux is

$$\mathbf{J}_w = \mathbf{J}_a + \mathbf{J}_d = \mathbf{v}(w + P) - \kappa \nabla T, \tag{3.18}$$

and the source/sink term is

$$g = g_J + g_{\text{damp}} = \mathbf{E} \cdot \mathbf{J}_{\text{conv}} + \frac{w - w_0}{\tau_w}.$$

The equation, (3.12), is now immediate from (3.14), if the specified identifications are made.

3.2.2 The Momentum Subsystem

The approach here is somewhat different. We follow the lead of [93]. We take as starting point the formulation,

$$n \frac{d\mathbf{v}}{dt} = -\nabla P + n\mathbf{F}, \tag{3.19}$$

of Newton's second law in a coordinate system moving with the fluid, a so-called inertial coordinate system. The symbol P retains its previous meaning as the specific pressure, and $\mathbf{F}$ denotes the "body forces" in a unit mass acting on the fluid particles. The left hand side of (3.19) is the product of the particle density with the acceleration in inertial coordinates.

Now the chain rule of calculus gives,

$$\frac{d\mathbf{v}}{dt} = \frac{\partial \mathbf{v}}{\partial t} + \mathbf{v} \cdot \nabla \mathbf{v},$$

when the acceleration is expressed with respect to a fixed coordinate system. We get, then, for Newton's law of motion in a fixed coordinate system,

$$n \left(\frac{\partial \mathbf{v}}{\partial t} + \mathbf{v} \cdot \nabla \mathbf{v} \right) = -\nabla P + n\mathbf{F}. \tag{3.20}$$

If we add the left hand side of the particle conservation equation, (3.10), multiplied by $\mathbf{v}$, to the left hand side of (3.20), we obtain the vector equation,

$$\frac{\partial (n\mathbf{v})}{\partial t} + \mathbf{v} \cdot \nabla (n\mathbf{v}) = -\nabla P + n\mathbf{F}.$$

It remains to identify the force function, $\mathbf{F}$. Here, this is represented by an electrical force per unit mass,

$$\mathbf{F}_{\text{elect}} = \frac{e}{m}\mathbf{E},$$

and a resistive force per unit mass,

$$\mathbf{F}_{\text{damp}} = -\frac{\mathbf{v}}{\tau_p},$$

which is a generalization of Stokes' law. Note that $\frac{1}{\tau_p}$ has the dimensions of a frequency, and this is to be interpreted in terms of frequency of collisions, with the resistive force proportional to velocity. Altogether, we obtain (3.11) from (3.20), with $\mathbf{F}$ defined as the sum of $\mathbf{F}_{\text{elect}}$ and $\mathbf{F}_{\text{damp}}$. It is shown in [50, p. 51] how to derive the electrical force from electrical surface tension arguments, via the Maxwell tensor. In this approach, derived in a so-called laboratory frame, the electrical force does not arise as a source term at all, but as the divergence of the Maxwell tensor.

3.3 Subsonic Linearization Analysis in One Dimension

In this section, we order the basic variables as v, n, T, and ϕ, because of symmetry considerations related to the variables v and n. The physical domain is the interval $\bar{G} = [a, b]$. Dirichlet boundary conditions are imposed, on n, T, ϕ, with $n(a) = n(b)$. In one dimension, the steady-state equations are obtained from (3.9–3.12) by setting the time derivatives equal to zero. Since, by the conservation of mass equation, $nv =$ const. holds, and since the boundary conditions for n agree at the endpoints, it follows that the boundary conditions for v are periodic (but not specified). The map, whose action is defined on the basic variables by bringing all nonzero terms to the left hand side of the steady-state version of system (3.9–3.12), and by dividing (3.11) by mn, and (3.12) by n, is called Φ. The domain of Φ is made precise below. The linearized equations, describing the Fréchet derivative of Φ, thus assume the following form, where the boundary conditions are homogeneous Dirichlet conditions for δn, δT, and $\delta\phi$, and the boundary conditions for δv are prescribed to be periodic,

$$\begin{bmatrix} 0 \\ 0 \\ -\frac{1}{n}\frac{d}{dx}(\kappa\frac{d\delta T}{dx}) \\ -\epsilon\frac{d^2\delta\phi}{dx^2} \end{bmatrix} + \begin{bmatrix} A & B \\ C & D \end{bmatrix} \frac{d}{dx} \begin{bmatrix} \delta v \\ \delta n \\ \delta T \\ \delta\phi \end{bmatrix} + \begin{bmatrix} E & F \\ G & H \end{bmatrix} \begin{bmatrix} \delta v \\ \delta n \\ \delta T \\ \delta\phi \end{bmatrix} = f. \tag{3.21}$$

The (spatially dependent) eigenvalues of the *symmetric* matrix A are calculated to be

$$\lambda = \frac{1}{2}\left(n + \frac{kT}{mn}\right) \pm \frac{1}{2}\sqrt{\left(n + \frac{kT}{mn}\right)^2 - 4\left(\frac{kT}{m} - v^2\right)}. \tag{3.22}$$

Here,

$$A = \begin{bmatrix} n & v \\ v & \frac{kT}{mn} \end{bmatrix}, \tag{3.23}$$

and the smaller eigenvalue is positive if n and T are strictly positive, and if

$$v^2 < \frac{kT}{m} = c^2. \tag{3.24}$$

This type of point in function space is termed a *subsonic* point. This case was first considered in [47], where damped Newton/standard finite difference methods were presented. When Newton's method is employed in this way, it is essential to determine conditions under which the linear increments are appropriately bounded in a neighborhood of a (potential) root of Φ. This is equivalent to uniform bounds for the operator derivative inverse maps, and represents one of the three properties for an (exact) operator Newton method to yield existence of a root, and R-quadratic convergence. The latter means that the error is bounded by an expression, of the form,

$$\rho^{2^k},\ k \to \infty,\ \rho < 1.$$

The other two requirements are sufficient regularity, and a sufficiently small starting residual, as measured in the range space norm (cf. [66]).

Explicit representations of B, C, D and of E, F, G, H may be given as follows:

$$B = \begin{bmatrix} 0 & 0 \\ \frac{k}{m} & \frac{-e}{m} \end{bmatrix},\ C = \begin{bmatrix} 3w - 2kT & \frac{v}{n}(w + kT) - \frac{\kappa}{n^2}\frac{dT}{dx} \\ 0 & 0 \end{bmatrix}, \tag{3.25}$$

$$D = \begin{bmatrix} \frac{5}{2}kv & -ev \\ 0 & 0 \end{bmatrix},\ E = \begin{bmatrix} \frac{dn}{dx} & \frac{dv}{dx} \\ \frac{dv}{dx} + \frac{1}{\tau_p} & \left(\frac{-k}{mn^2}\right) T \frac{dn}{dx} \end{bmatrix}, \tag{3.26}$$

$$F = \begin{bmatrix} 0 & 0 \\ \frac{k}{mn}\frac{dn}{dx} - \frac{v}{\tau_p^2}\tau'_p & 0 \end{bmatrix},\ G = \begin{bmatrix} g_{11} & g_{12} \\ 0 & 0 \end{bmatrix}, \tag{3.27}$$

$$H = \begin{bmatrix} \frac{5}{2}k\frac{dv}{dx} + \frac{5v}{2n}k\frac{dn}{dx} + \frac{\frac{dw}{dT}\tau_w - (w - w_0)\tau'_w}{\tau_w^2} & 0 \\ 0 & 0 \end{bmatrix}, \tag{3.28}$$

where $\tau'_p = \frac{d\tau_p}{dT}$, $\tau'_w = \frac{d\tau_w}{dT}$, and

$$g_{11} = 3mv\frac{dv}{dx} + \frac{5}{2}k\frac{dT}{dx} + \frac{5}{2n}kT\frac{dn}{dx} + \frac{3}{2n}mv^2\frac{dn}{dx} - e\phi' + \frac{1}{\tau_w}\frac{dw}{dv}, \tag{3.29}$$

$$g_{12} = -\frac{v}{n^2}(w + kT)\frac{dn}{dx} + \frac{1}{n^2}\frac{d\kappa}{dx}\frac{dT}{dx}. \tag{3.30}$$

The derivative entries of E, and the associated requirement that these entries be pointwise bounded in the estimation, suggests that n and v be taken from $W^1_\infty(G)$, the space of measurable functions with essentially bounded derivatives. To determine the corresponding spaces for T and ϕ, we note that the choice of the function space is not uniquely determined. We select $W^2_\infty(G)$, the integral of $W^1_\infty(G)$, as the appropriate space for T and ϕ. Thus, the system map Φ, formally defined as above, including the boundary conditions of the model, accordingly has the domain $D_\Phi \subset X = \prod_1^2 W^1_\infty \times \prod_1^2 W^2_\infty$, and range in $Y = \prod_1^4 L_\infty$, *and* (3.24) will hold for every element in a closed ball $B_{r_0} \subset X$, centered at a subsonic point $u_0 \in X$, such that n and T are strictly positive, *if* r_0 is sufficiently small. It is appropriate to assume at the outset, then, that $D_\Phi \subset B_{r_0}$, so that every function point in D_Φ satisfies (3.24); we may also assume that n and T are uniformly bounded away from zero in this set.

The Lipschitz property of the map Φ', where here we view the system (3.21) as the representation for $\Phi'(v, n, T, \phi)(z, \omega) = f$, with

$$z = (\delta v, \delta n), \qquad \omega = (\delta T, \delta\phi), \qquad f = (f_1, f_2), \tag{3.31}$$

is evident from the representation for Φ'.

The uniform inverse bounding proceeds as follows. As shown in [47], with the function spaces selected in this paper, the H^1 product norm of z can be estimated in terms of the L_2 norms of ω, ω', and f_1, under the conjunction of the hypothesis (3.24) and the $L_2 \times L_2$ coerciveness *assumption*,

$$\Lambda = E + E^* - \frac{dA}{dx} \text{ is uniformly positive definite.} \tag{3.32}$$

Here, E^* is the matrix transpose of E. The boundedness argument uses the fact that the auxiliary operator,

$$Jz = Az' + Ez,$$

has the same domain boundary conditions as its adjoint, J^*, permitting the inner product relation,

$$(Jz, z) = \frac{1}{2}\left[(Jz, z) + (J^*z, z)\right] = \left(\frac{1}{2}\Lambda z, z\right).$$

A final calculation, making use of the inner product of $[0, \omega]$ with (3.21), and making use of the *hypothesis*,

$$h_{11} \text{ is positive, and sufficiently large,} \tag{3.33}$$

where h_{11} is the nonzero entry of H, shows that the H^1 product norm of $[z, \omega]$ is estimated in terms of the L_2 norm of f. This series of calculations controls the L_∞ norm of $[z, \omega]$; the L_∞ norm of $[z', \omega'']$ is now estimated by direct use of the system (3.21), making use of the fact that $[v, n, T, \phi] \in B_{r_0}$. We have now outlined the proof of the following (see also §6.3).

Theorem 3.3.1. *Let $\bar{G} = [a, b]$, and let the function spaces X and Y be selected as above, Let the steady-state system map Φ be given with Dirichlet boundary conditions on n, T, ϕ, and periodic boundary conditions on v, such that $D_\Phi \subset B_{r_0}$, where every point in B_{r_0} is a subsonic point, with uniform positivity bounds on n and T. If* (3.32) *and* (3.33) *hold, and x_0 is such that $\Phi(x_0)$ is sufficiently small, then an R-quadratically convergent Newton sequence $\{x_k\}$ may be defined in the standard way, with limit x, satisfying $\Phi(x) = 0$.*

3.4 Energy Transport Models and Stokes' Flow

In fluid mechanics, a Stokes' flow is one in which inertial effects are assumed negligible, in the sense that the product of the velocity and velocity gradient is small compared to the pressure gradient–density ratio. Energy transport models are of this type. Specifically, if $\nabla\mathbf{v}$ denotes a matrix induced by column/row "multiplication", then

$$|\mathbf{v}\cdot \nabla\mathbf{v}| << \left|\frac{\mathbf{v}}{\tau_p}\right| \tag{3.34}$$

characterizes such models. We shall consider steady-state models, involving single carrier transport. Note that the relation (3.34) is implied by

$$\tau_p\|\nabla\mathbf{v}\| << 1 \text{ (Stokes' flow).} \tag{3.35}$$

When (3.35) holds, called the inertial approximation, there are two fundamentally independent ways to proceed.

I Either we can proceed directly from the hydrodynamic equations, (3.10, 3.11), and the inertial approximation to obtain a constitutive relation for the current density in terms of potential, concentration, and temperature gradients (see §3.5 for an outline of this approach); or,

II The *microscopic* relaxation time approximation can be employed, as distinct from the macroscopic approximation above.

In either case, the three dependent variables of the problem are electrostatic potential, ϕ, carrier concentration, n, and equivalent carrier temperature, T. Although models of this type are widely employed in the engineering community, it is relatively recent that such models have been derived via microscopic relaxation time approximations. An example of the latter class of models, which have proven highly effective in simulation, will now be presented. The model was introduced in [28], in an effort to represent nonlocal mobility functions, while retaining essential structural features of the drift-diffusion model. The equations, unlike those of the hydrodynamic model, do not possess hyperbolic modes. The model was first analyzed in [77]. It should be noted that the model presented here differs from the simple reduction of the hydrodynamic model via the inertial approximation (3.35), since nonparabolic energy band structure and a quadratic (not linear) energy–temperature relation are incorporated.

3.4.1 The Steady-State System

We shall present only the isotropic version of this model. The system is given as follows.

$$-\nabla\cdot(\epsilon\nabla\phi) + en = ek_1, \tag{3.36}$$

$$\nabla\cdot\mathbf{J} = 0, \tag{3.37}$$

$$\nabla\cdot\mathbf{S} = -\nabla\phi\cdot\mathbf{J} - nC_{col}. \tag{3.38}$$

Here, $\mathbf{J}$ and $\mathbf{S}$ represent the current density and energy density, respectively, and C_{col} is a collision rate. They are given by the expressions,

$$\mathbf{J} = e\left\{-\mu n\nabla\phi + \frac{k}{e}\nabla(\mu nT)\right\}, \tag{3.39}$$

$$\mathbf{S} = -Q(T)\left\{\frac{k^2}{e}\nabla(\mu nT^2) - \mu nkT\nabla\phi\right\}, \tag{3.40}$$

$$C_{col} = c_1 e\left\{\left(1+\frac{1}{2}T\right)T - T_0\right\}, \tag{3.41}$$

with the mobility μ and the effective flow factor $Q(T)$ functions of temperature; the latter function is assumed to preserve uniform ellipticity. Typical representations are:

$$\mu = \mu_0 T_0/T, \tag{3.42}$$

$$Q(T) = \frac{3}{2}\left(1 - \alpha kT/2\right). \tag{3.43}$$

Here, α is a nonparabolic-band factor, so that in the modeling leading up to these equations, microscopic carrier kinetic energy is given by $E(1+\alpha E)$, where E is microscopic carrier energy. Also, μ_0 is a low field mobility constant, k is Boltzmann's constant, T_0 is the lattice temperature, c_1 is a numerical constant, e is a charge modulus constant, and ϵ is the dielectric constant. The function k_1 is the *nonnegative* spatially dependent doping function. Boundary conditions are mixed Dirichlet/Neumann boundary conditions; on a one-dimensional device, these are Dirichlet endpoint conditions.

3.4.2 Exponential Variables in One Dimension

In this subsection, we introduce a set of exponential variables for the domain, $\bar{G} = [a, b]$. They will be used in the next subsection to deduce corresponding maximum principles, which facilitate the definition of an invariant map. The fixed points coincide with the system solutions.

The exponential variables are similar to the quasi-Fermi levels of the drift-diffusion model. They are determined by implicit relations involving the variables ϕ, n, and nT of the old set; ϕ is the first variable of the new set. The variables v and z of the new set are implicitly given. Also, it is critical to interpret μ as an explicitly given (composite) function in what follows.

$$n(x) = \exp(c_0[\Phi(x) - v(x)]), \tag{3.44}$$

$$n(x)T(x) = \exp(c_0[\Phi(x) - z(x)]), \tag{3.45}$$

where

$$\Phi(x) = \int_a^x \mu(t)\phi'(t)\, dt. \tag{3.46}$$

We shall now rewrite the system in terms of these variables. Although μ must be explicit, the system is still strictly logically equivalent to the two systems defined in the previous section. We shall actually make use of the exponential variables when lagging has been employed in the original system as a means of defining the fixed point map.

$$-(d/dx)(\epsilon d\phi/dx) + e\exp([\Phi - v]c_0) = ek_1, \tag{3.47}$$

$$(d/dx)(\exp([\Phi - v]c_0)dv/dx) = 0, \tag{3.48}$$

$$(d/dx)(kQ(T)\exp([\Phi - z]c_0)dz/dx) = -d\phi/dx\, J - F(\Phi, v, z), \tag{3.49}$$

$$F(\Phi, v, z) = c_1 e \exp(c_0\Phi)\{(1 + (1/2)\exp([v - z]c_0))\exp(-zc_0) - T_0\exp(-vc_0)\}. \tag{3.50}$$

Note that in this system, $c_0 = \frac{e}{kT_0\mu_0}$, and that there is a simplified formula for the constant J: $J = -ev'n$, so that v' may be identified with carrier velocity. We agree to track right-left current flow, i.e., $v' > 0$ and $J < 0$. The equation (3.48) is an example of Bernoulli's equation, with exponent 2, for v'. The solution is given by,

$$v'(x) = Y(x)\frac{1}{C - c_0\int_a^x Y(t)\, dt}, \tag{3.51}$$

where

$$Y(x) = \exp(-c_0\Phi(x)), \quad C = \frac{1}{v'(a)} > 0. \tag{3.52}$$

We close this subsection with a relation for J.

Proposition 3.4.1. *The constant current J is negative and satisfies*

$$J = e \exp(-c_0 v(x_*)) \frac{v(a) - v(b)}{(b-a)} / \int_a^b Y(x)\, dx, \tag{3.53}$$

for some $x_* \in (a, b)$.

Proof. Since J is constant, it may be evaluated at any x:

$$J = e \frac{\exp(-c_0 v(x))}{c_0 \int_a^x Y(t)\, dt - C}. \tag{3.54}$$

The appropriate value is $x = x_*$, where x_* is guaranteed, by the mean value theorem of integral calculus, applied to $\int_a^b v'(x)\, dx$, to satisfy

$$c_0 \int_a^{x_*} Y(x)\, dx - C = \frac{\int_a^b Y(x)\, dx}{v(a) - v(b)}. \tag{3.55}$$

The proposition follows from these relations.

3.4.3 Maximum Principles

The key to defining a fixed point map, $\mathbf{T}$, on the variables v and z is the establishment of maximum principles, which induce a mapping invariance property. In this section, we derive these necessary results. We begin by describing the map intuitively; ϕ is computed as a fractional step, if initial values of v and z are prescribed. This employs $\mu(v, z)$, as a means of determining Φ. System decoupling proceeds as follows. In (3.48), Φ and v are substituted in the exponential; in (3.49), Φ and v are substituted throughout, but z is substituted in $Q(T)$ only. New values, v^* and z^*, and hence the image coordinates of the map, are then computed via the decoupling.

If boundary values of v are chosen so that $v(a) < v(b)$, then the discussion of the previous subsection shows that $v' > 0$, hence v satisfies the elementary maximum principle,

$$v(a) \leq v \leq v(b). \tag{3.56}$$

A separate argument shows this to be the case if $v(a) = v(b)$. Prior to passing to a discussion of boundary values for z, we note that the specifications,

$$\phi(a) \leq \phi(b), \quad n(b) \leq n(a), \tag{3.57}$$

imply $v(a) \leq v(b)$, hence, (3.56). Bounds for z are considered next. The possible change of sign of ϕ' makes standard approaches inapplicable. We begin as follows. Here, $t^- := \min\{t, 0\}$.

Lemma 3.4.1. *Let* $\gamma = \min(0, \gamma_*)$, *where* $\exp(c_0\gamma_* - c_0 v(a)) - \inf k_1 = 0$. *If Y is the function defined in* (3.52), *then* $\psi = (Y^{-1} - \exp(c_0\gamma))^- \equiv 0$ *if* (3.57) *holds. In particular,* $Y^{-1} \geq \exp(c_0\gamma)$ *and* Φ *of* (3.47) *satisfies, without restriction on z,*

$$\Phi \geq \Phi_{\min} := \gamma. \tag{3.58}$$

Proof. Clearly, $\psi(a) = 0$. The assumption, (3.57), guarantees that $\Phi(b) > 0$, hence $Y^{-1}(b) > 1$, so that $\psi(b) = 0$. We shall employ ψ as a test function in the weak formulation of (3.47). Now one sees that the function,

$$en - ek_1 = e(\exp([\Phi - v]c_0) - k_1), \tag{3.59}$$

arising in (3.47), is nonpositive on the set, $\{\Phi \le \gamma_*\}$. The integrated product of this function with ψ is therefore nonnegative. Moreover,

$$\int_a^b \epsilon\phi'\psi'\,dx = \int_{\{\Phi\le\gamma\}} \frac{\epsilon Y}{\mu c_0}|\psi'(x)|^2\,dx \ge 0,$$

so that, from the weak relation, the latter integral vanishes. We conclude that $\psi \equiv 0$, and the lemma follows.

Definition 3.4.1 (roots). *We assume the existence of a root, $z = z_{**}$, satisfying*

$$\begin{aligned} 0 &= c_1 e \exp(c_0\Phi_{\min})[\{1 + (1/2)\exp(c_0[v(b) - z_{**}])\}\exp(-c_0 z_{**}) \\ &- T_0\exp(-c_0 v(b))] + \frac{e^2 \sup k_1}{\epsilon Y_{\min}}\frac{v(b) - v(a)}{b - a}\exp(-c_0 v(a)), \end{aligned} \tag{3.60}$$

where

$$Y_{\min} = \exp[-c_0(\phi(b) - \phi(a))\mu_0 T_0 \exp(z_{**} - v(a))]. \tag{3.61}$$

Moreover, we select the root, $z = z_*$, satisfying

$$\begin{aligned} 0 &= c_1 e \exp(c_0\Phi_{\min})[\{1 + (1/2)\exp(c_0[v(a) - z_*])\}\exp(-c_0 z_*) \\ &- T_0\exp(-c_0 v(a))] - \frac{e^2 \sup k_1}{\epsilon Y_{\min}}\frac{v(b) - v(a)}{b - a}\exp(-c_0 v(a)). \end{aligned} \tag{3.62}$$

Finally, we define

$$z_{\max} = \max(z(a), z(b), z_{**}),\quad z_{\min} = \min(z(a), z(b), z_*).$$

Proposition 3.4.2. *Suppose that the solution ϕ of (3.47) satisfies*

$$\phi'(a) \ge 0,\quad \phi'(b) \le 0, \tag{3.63}$$

for $v(a) \le v \le v(b)$ and $z_{\min} \le z \le z_{\max}$. Then the bound,

$$|\phi'| \le \frac{e(b-a)}{\epsilon}\sup k_1, \tag{3.64}$$

holds, and the function F of (3.50) satisfies

$$F(\Phi, v, z) + \phi' J \;\le\; 0,\quad z \ge z_{\max}, \tag{3.65}$$

$$F(\Phi, v, z) + \phi' J \;\ge\; 0,\quad z \le z_{\min}. \tag{3.66}$$

In particular, for solutions of (3.49), the maximum principle follows:

$$z_{\min} \le z \le z_{\max}. \tag{3.67}$$

Proof. The bound (3.64) follows directly from (3.47) and the hypotheses, (3.63). The relations (3.53) and (3.64), and the definition of F in (3.50), lead to (3.65) and (3.66). The maximum principle for (3.49) follows as in the proof of Lemma 3.4.1, via the respective choices of positive and negative parts for test functions, $(z - z_{\max})^+$, $(z - z_{\min})^-$, yielding upper and lower bounds.

3.4.4 The Fixed Point Map

We shall prove that a solution exists in this subsection.

Theorem 3.4.1. *Let $G = (a, b)$, and suppose (3.60) and (3.63) hold. Set K equal to the closed convex subset of $\prod_1^2 L_2(G)$ defined by,*

$$K = \{[v, z] : v(a) \le v \le v(b), \ \ z_{\min} \le z \le z_{\max}\}. \tag{3.68}$$

Let $\mathbf{T}$ be the mapping, invariant on K, introduced in §3.4.3. Then $\mathbf{T}$ is well-defined, acts continuously on K, and has relatively compact range. In particular, $\mathbf{T}$ has a fixed point, $[v, z]$, and the triple, $[\phi(v, z), v, z]$, defines a solution of the steady-state system, (3.36–3.38).

Proof. The fact that $\mathbf{T}$ is well-defined follows from quadratic minimization applied to the functional associated with (3.48) and convex minimization applied to (3.49); note that F is decreasing in z. These are individual gradient equations. The details of a similar argument may be found in [67] or in the following chapter. Here, a slight modification is required for the determination of ϕ, as computed in the fractional step. What is necessary is a preliminary mapping of $\phi \mapsto \Phi$. The relative compactness follows from H^1 bounds for v^* and z^*; test functions, $v^* - v_I$ and $z^* - z_I$, are employed, where the linear interpolants of the boundary values are subtracted. The routine estimates here employ the maximum principles.

Continuity is established for v^* and z^* by using the components of the difference of two solutions as test functions: $v_2^* - v_1^*$, $z_2^* - z_1^*$. The preliminary continuity of the dependence of ϕ must first be established, however. We illustrate this. Starting with the identity,

$$\phi_2'' - \phi_1'' = \frac{e}{\epsilon}\left(\exp([\Phi_2 - v_2]/c_0) - \exp([\Phi_1 - v_1]/c_0)\right), \tag{3.69}$$

we employ $\psi = \phi_2 - \phi_1$ as test function. This yields, after some estimation, constants C_1 and C_2 such that

$$\|\psi'\|_{L_2}^2 \le C_1 \|\psi\|_{L_2}^2 + C_2 \|\psi\|_{L_2} (\|\mu_2 - \mu_1\|_{L_2} + \|v_2 - v_1\|_{L_2}). \tag{3.70}$$

Here, we have used previously derived estimates for ϕ', as well as the maximum principles and elementary calculus. Though the positive function μ, in the definition of Φ, need not be continuous, it can be approximated by such, and thus the mean value theorem of integral calculus may be employed. The Poincaré inequality in one dimension, applied to ψ, allows us to conclude the continuous H^1 dependence of ϕ upon the L_2 variation of v and z, if $b-a$ is sufficiently small; in this case, the terms involving $\|\psi\|_{L_2}$, on the right hand side of (3.70), can be absorbed on the left hand side. However, a simple change of variable then gives the general result. The arguments for the continuity of the other variables follow closely in spirit Lemmas 3.3 and 3.4 of [67]. An application of the Schauder fixed point theorem then yields a fixed point. This can be identified with either set of dependent variables.

Remark 3.4.1. The condition (3.63) appears to hold approximately, on the basis of numerical simulations, carried out on $n^+ - n - n^+$ diodes. A so-called neutral region surrounds the contacts, and the field is approximately zero. See [77] for further details.

3.5 Modeling Issues

Two intriguing aspects of the hydrodynamic model have emerged since its study began intensively in the middle 1980s. One of these, first documented in [49], relates to the sensitivity of the model to the heat conduction term. The other, first observed in [44], deals with the possibility of shock formation for certain parameter ranges of the model. When the Wiedemann-Franz law is used to model κ, so that

$$\kappa = \frac{3}{2}\frac{\mu_0 k^2}{e} n T_0, \tag{3.71}$$

where μ_0 denotes the low field mobility, it was discovered that a spurious velocity overshoot occurs near the channel-drain junction in an n^+-n-n^+ diode. Empirical modification of (3.71), whereby 1.5 was replaced with .4, was shown to bring the velocity into more consistent alignment with the data of Monte-Carlo simulations of the Boltzmann transport equation. This is discussed at length in [75], where comparisons are presented. The development of the energy transport model was significantly motivated by the goal of rectifying this seemingly arbitrary aspect of the hydrodynamic model; simulations reveal no spurious overshoot in the n^+-n-n^+ diode. Alternatively, the authors of [126] introduced a modified heat flux, based on first principles, into the hydrodynamic model. It has been found, however, that the factor most affecting the spurious overshoot is the low field mobility; when chosen to be doping (spatially) dependent, the overshoot is reduced. This fact is documented in [78], which uses the hyperbolic, explicit, time-stepping algorithm ENO (essentially nonoscillatory), developed in [123]; it had earlier been applied in [37].

A realistic case study of shock regimes has been given in [45]. Earlier, independent numerical methods were employed in [36] to identify length and temperature regimes in which shocks are likely to occur. A major question, which must be dealt with if the full convective hydrodynamic model becomes a mainstay of the device simulation community, is the identification of the appropriate parameter ranges in which devices actually function. This dramatically affects the choice of algorithm; shock capturing algorithms, such as ENO or RKDG (Runge Kutta Discontinuous Galerkin), are suitable in regimes of shocks and steep fronts, and where convection dominates diffusion and reaction terms. A comparison of ENO and RKDG was, in fact, carried out in [27]. On the other hand, if the regime is subsonic, and the steady-state system is thereby elliptic, it would be appropriate to employ standard exponential fitting in discretizations. It is known that the inertial approximation leads to elliptic steady-state systems. We study this a little more closely.

3.5.1 Regimes Defined by Damping

We identify three essential regimes defined by the hydrodynamic model:

1. The full hydrodynamic regime;
2. A restricted limiting regime, in which inertial effects are negligible over the momentum relaxation time scale;
3. A fully restricted limiting regime, contained within the preceding, in which temperature variation is negligible, so that heat flux is insignificant. This corresponds to the drift-diffusion regime. It is characterized as friction dominated.

In the first regime, friction is not dominant. In the second, resistance to momentum transitions is significant over the relaxation time scale. In the third regime, additional resistance to heat flow (temperature gradients) is significant. Regime two may be defined via a critical limit of the hydrodynamic model, as we now explain. The equations (3.9, 3.10) and (3.12) remain as in the hydrodynamic model. However, (3.11) is rewritten, as in [116], to yield:

$$\mathbf{J} + \tau_p(\mathbf{J} \cdot \nabla)\mathbf{v} = eD\left[\frac{T}{T_0}\nabla n + n\nabla\left(\frac{T}{T_0} - \frac{e}{kT_0}\phi\right)\right]. \tag{3.72}$$

Here, D is the diffusion coefficient. To obtain the second regime, we formally take the limit,

$$\tau_p \to 0,$$

to obtain a constitutive relation for the current, given by:

$$\mathbf{J} = eD\left[\frac{T}{T_0}\nabla n + n\nabla\left(\frac{T}{T_0} - \frac{e}{kT_0}\phi\right)\right]. \tag{3.73}$$

This is the interpretation of the inertial approximation . The third regime follows from the second when isothermal conditions prevail. A final regime, not singled out explicitly, is the isentropic regime. It is expected to be present for small values of the constant c_w.

In a paper [76] presented at the May, 1994, Portland Workshop on Computational Electronics, the validity of the inertial approximation (regime two) was analyzed in the context of the n^+-n-n^+ diode and the MESFET transistor. This has been a widely accepted approximation in the device community, particularly since it allows for algorithms based on exponential fitting, inasmuch as shocks cannot occur in this case. The approximation was shown to be invalid in the vicinity of diode junctions and MESFET contacts, while holding quite reliably elsewhere. What is clear is that the regimes not only allow definition parametrically, but spatially as well. Independent evidence suggests the strong dependence of the regimes on the magnitude of the constant c_w.

The conservation law subsystem of the hydrodynamic model is an example of an incompletely parabolic system. These have been studied for well-posedness (cf. [55, 124]). The theory roughly asserts that the parabolic and hyperbolic subsystems can be *conceptually* decoupled with respect to boundary conditions. The so-called inflow conditions are associated with the hyperbolic part. A steady-state one-dimensional model, with the energy equation replaced by an adiabatic constitutive law, has been rigorously analyzed in [41]. This is the isentropic case. The corresponding evolution system has been rigorously analyzed in [139]. Gamba has analyzed the full steady-state model, subject only to certain assumptions about the existence of invariant regions, in [43]. We shall close this chapter by describing a promising new model which allows for quantum transport.

3.6 A Glimpse of the Quantum Hydrodynamic Model

Ultrasmall devices involve critical lengths approaching the coherence length, or inelastic mean free path, of the electron carriers. Ensuing quantum effects modify the electron and current density distributions. Modifications of the HD model are certainly necessary to incorporate these effects, as shall be documented later. The quantum hydrodynamic (QHD) model is a moment model, derived from the Wigner equation. The Wigner density function may be defined as a function of position and momentum by the transform,

$$w(\mathbf{x},\mathbf{p}) = \frac{1}{(2\pi\hbar)^n}\int_{R^n} \rho\left(\mathbf{x}+\frac{\hbar\omega}{2},\mathbf{x}-\frac{\hbar\omega}{2}\right) e^{i\mathbf{p}\omega}\, d\omega, \tag{3.74}$$

where ρ (the analog of the function f in the BTE) is the density "matrix", the kernel of the statistical operator $\mathbf{S}$; the latter depends upon the statistical distribution assumed for the particles, i.e. , upon $F(E,\beta = 1/T)$, where E is the energy and T is the absolute scaled temperature. In terms of F, $\mathbf{S}$ may be expressed in the notation of the spectral calculus by

$$\mathbf{S}(\beta) = \mathbf{F}(\mathbf{H},\beta), \tag{3.75}$$

where $\mathbf{H}$ is the Hamiltonian operator appearing in the single particle, effective mass Schrödinger equation. The self-adjointness of the associated operators has been derived in [125] for F defining both Boltzmann statistics and Fermi-Dirac statistics.

The Wigner density function satisfies the collisionless time evolution Wigner equation,

$$\frac{\partial w}{\partial t} + \frac{\mathbf{p}}{m} \cdot \nabla_{\mathbf{x}} w + \theta f = 0, \tag{3.76}$$

where θ is a nonlocal operator, making use of the self-consistent potential in its kernel. A derivation of (3.76) can be based on the Schrödinger, or (equivalent) Heisenberg, equation (see [101]). A derivation of the general QHD model is carried out by Gardner in [46], making use of Wigner's approximate solution of (3.76) (see [138]). Gardner also employs relaxation expressions similar to those of the classical theory. The QHD model may be viewed as a quantum corrected version of the classical hydrodynamic equations, with the pressure tensor and the energy density corrected by $O(\hbar^2)$ perturbations. These are represented by:

$$P_{ij} = nkT\delta_{ij} - \frac{\hbar^2 n}{12m} \frac{\partial^2}{\partial x_i \partial x_j} \log(n), \tag{3.77}$$

$$W = \frac{3}{2}nkT + \frac{1}{2}mnv^2 - \frac{\hbar^2 n}{24m} \nabla^2 \log(n). \tag{3.78}$$

Some applications of the QHD model, such as the resonant tunnel diode, also involve quantum well potentials. The expression for the quantum corrected pressure tensor was derived in [4, 5], and a one-dimensional version of the model was formulated in [53]. The model is of considerable physical and practical importance. An example of this is furnished by the regions of negative differential resistance detected in the voltage-current curve associated with the model; closely tied to this is an hysteresis effect, which has been confirmed computationally in [26]. Earlier, the authors of [88] had discovered this effect in the Wigner equation. The QHD model is computationally more tractable than the Wigner formulation, however. Thus, it is well suited to studying the multiple states arising in this model, with implications for logic and memory devices.

No complete existence theory is available for the QHD model for semiconductors, either in steady-state or in long time evolution. The quantum equivalent of the isentropic model, incorporating a quantum pressure tensor perturbation and an adiabatic pressure–density relation, has been recently analyzed by Zhang and the author in steady-state (cf. [140]). The quantum well case of a discontinuous potential was not included in this work. This reduced model had been discussed earlier in [53]. We present the basic equations as follows.

$$n_t + (nv)_x = 0, \tag{3.79}$$

$$(mnv)_t + (mnv^2 + p(n) + Q(n))_x = -n\phi_x - \frac{mnv}{\tau}, \tag{3.80}$$

$$\phi_{xx} = e(N_D - N_A - n), \tag{3.81}$$

where n is the electron density, v the velocity, and ϕ the electrostatic potential. The pressure function, $p = p(n)$, has the property that $n^2p'(n)$ is strictly monotonically increasing from $[0, \infty)$ onto $[0, \infty)$. A commonly–used hypothesis is:

$$p(n) = cn^{\gamma}, \qquad \gamma > 1, \qquad c > 0. \tag{3.82}$$

Quantum mechanics is represented by the quantum potential (see [46]):

$$Q(n) = -\frac{\hbar^2}{12m} n(\log(n))_{xx}. \tag{3.83}$$

The device domain is the x–interval, $G \equiv (0,1)$. The given functions N_D and N_A are the usual density of donors and the density of acceptors, respectively.

The approach used in [140] was completely novel to this application area, in that the system was reduced to an integro-differential equation, with a set of boundary conditions, including a nonstandard second order boundary condition, which is equivalent to specifying the quantum potential at the (current) inflow boundary. Indeed, for each specified (positive) value of the constant flux $j = nv$, and for $\omega = \sqrt{n}$, the latter is completely characterized by the integro-differential equation,

$$\begin{aligned} \frac{\hbar^2}{6m}\omega_{xx} &= \frac{mj^2}{2\omega^3} + \frac{c\gamma}{\gamma - 1}\omega^{2\gamma-1} + mj\omega \int_0^x \frac{dy}{\omega^2 \tau} \\ &\quad + \omega\left(e \int_0^1 G(x,\xi)(N_D - N_A - \omega^2)d\xi + x(\phi_1 - \phi_0) + b\right), \end{aligned} \tag{3.84}$$

$$\omega(0) = \omega_0, \quad \omega(1) = \omega_1, \tag{3.85}$$

where

$$b = \frac{\hbar^2}{12m}\frac{n_2}{n_0^2} - \frac{mj^2}{2n_0^2} - \frac{c\gamma}{\gamma - 1}n_0^{\gamma-1}, \ \omega_0 = \sqrt{n_0}, \ \omega_1 = \sqrt{n_1}.$$

If ω is a smooth solution of (3.84, 3.85), then

$$2\omega_0^3\omega_{xx}(0) = n_2.$$

The latter can be interpreted as a new quantum flux boundary condition which was discovered for mathematical well-posedness; n_0, n_1, ϕ_0, ϕ_1 are end-point boundary conditions on the electron density and the classical potential, and G denotes the Green's function for d^2/dx^2, taken to be *nonpositive*, while γ is the exponent in the pressure–density relation. 'A priori' estimates and an existence result were obtained via the Leray-Schauder fixed point theorem.

– As noted by Irene Gamba in a personal communication, the technique used to obtain the pointwise 'a priori' estimates is analogous to division by the dependent variable, and integration along the streamlines of the Bernoulli function. There is some hope, therefore, of eventually extending these results to two spatial dimensions. Moreover, it is likely that the distribution generalization of the QHD model, allowing for a quantum well, can be handled also. For readers unfamiliar with the quantum well, it represents a step discontinuity of the classical potential (redefined from the Poisson equation) over a physical portion of the device (the well), inducing distribution like behavior in the gradient, as it appears in the conservation equations. One such extension of the QHD model to this case has recently been carried out by Gardner and Ringhofer, where a cancellation of singularities in the momentum equation has been effected.

The quantum perturbation introduces a third order density perturbation in the momentum equation. The estimates of [140] clearly show that an $\hbar \to 0$ strong limit is not possible. This is reminiscent, for systems, of the oscillatory dispersion phenomenon noticed by the authors of [94]. Their fundamental model was an evolution equation, for which a weak limit was demonstrated and studied. It is not clear whether a weak limit can be demonstrated for this steady-state model; as noted earlier, the limiting equations do have a weak solution [41].

Part II

Computational Foundations

4. A Family of Solution Fixed Point Maps: Partial Decoupling

For the remainder of the book, we restrict our attention to the drift-diffusion model. Moreover, we employ the so-called quasi-Fermi levels, originally introduced in Chap. 2, and designated below by v and w. For our purposes, we shall thus study variants of the following model, discussed in §2.5.

Definition 4.0.1 (quasi-Fermi level formulation). *The drift-diffusion model is described by the system,*

$$-\epsilon\nabla^2 u + e^{u-v} - e^{w-u} = k_1, \quad \text{on } G, \tag{4.1}$$

$$-\nabla\cdot(\mu_n e^{u-v}\nabla v) - R(u,v,w) = 0, \quad \text{on } G, \tag{4.2}$$

$$-\nabla\cdot(\mu_p e^{w-u}\nabla w) + R(u,v,w) = 0, \quad \text{on } G, \tag{4.3}$$

subject to boundary conditions for u, v, and w. In fact, if $\bar{u}$, $\bar{v}$, and $\bar{w}$ are $H^1(G)$ functions, with restrictions to ∂G in $L_\infty(\partial G)$, then any solution triple of this system is required to have boundary trace agreeing with that of $(\bar{u},\bar{v},\bar{w})$ on the Dirichlet boundary, $\Sigma_D \subset \partial G$; homogeneous Neumann boundary conditions are required in a weak sense on the complement, Σ_N. The system (4.1–4.3) is to be interpreted in the weak sense. Here k_1 describes the doping in the device in intrinsic concentration units, and ϵ is the dielectric function; both are pointwise bounded, with the latter bounded below by a positive constant. R is the recombination function; the minimal assumption on R is contained in (4.70) below. Finally, in formulating this system, units have been selected so that $\frac{kT_0}{e} = 1$.

In those instances when the solution fixed point map, characterized as a Gummel decoupling map, is a strict contraction, considerable simplification results: uniqueness, and the convergence of successive approximations to the solution. We shall begin this chapter with a study of a simplified model, for which we can carry out precise estimation of the Lipschitz constant of the map. Strict contractiveness means that this constant is strictly less than one. This presentation, contained in §4.1, is based upon [68], presented at a SIAM-IEEE meeting in 1984. The ideas and techniques of the remainder of the chapter represent substantial improvements of the methods described by the author in [67].

A significant feature is the systematic use of weighted Sobolev spaces to handle the breakdown of uniform ellipticity in the current continuity equations, due to the form of the mobility coefficients, and their reciprocal dependence upon the modulus of the electric field. Since the Einstein relations are employed, this equivalently means that the diffusion coefficients vanish at the transition points on the boundary of G, where a transition point is defined to be common to the (relative) closures of the sets Σ_D and Σ_N, corresponding to Dirichlet and Neumann boundary conditions. According to regularity theory, this is a point where the electric field may become unbounded, and where the mobility and diffusion coefficients may thereby vanish. It is not our intention to engage here in the debate over whether this can actually occur in a physical device, and the associated suitability of the model in this case. We have chosen to take the model as presented, and, in the absence of *compelling* alternatives, analyze the mathematical issues. Note that one does not have 'a priori' knowledge of what the threshold is for a given model!

A second feature is the development and use of a family of fixed point maps, and the study of their properties. Existence is a by-product of this study, which attempts to validate various decoupling approaches to the system for purposes of theoretical computation. This has led to allowing coupled current continuity subsystems in the definition of the fixed point map, and associating with the current continuity subsystem, as formulated in terms of the quasi-Fermi levels, a well-defined map $\mathbf{VW}_f$, as described fully in §4.4. Uses of this map elsewhere entail the special case when the map can be written as

$$\mathbf{VW}_f = [\mathbf{V}_f, \mathbf{W}_f].$$

In this case, the recombination term is necessarily taken to be zero. This is the situation considered in Chap. 5, and, in the context of the Slotboom variables, in §4.1. In the general case, the approach we take to the construction of (the family) $\mathbf{VW}_f$, which defines a family $\mathbf{T} = \mathbf{T}_f$ through the composition $\mathbf{T}_f = \mathbf{VW}_f \circ \mathbf{U}_f$, involves lagging the recombination term. The verification that this procedure yields a well-posed system follows by characterizing the maximum principles for the decoupled current continuity subsystem as obstacles or constraints. Obstacle problems, in turn, are naturally treated by variational inequalities, and our study proceeds in this way, culminating in the decisive finding that the constraints are inactive as discussed in §4.6: the variational inequality is equivalent to a system, defined in weak form with appropriate test functions.

The existence result for the complete drift-diffusion system is contained in §4.7.3. The reader should realize that, if the limited goal of existence is all that is desired, then the particular Gummel map may be selected through complete lagging of the variables v and w in F (cf. (4.70)), with the exponential term unlagged. If one is willing to add the hypothesis that R is decreasing in v and increasing in w, not made here, then complete system decoupling is possible, thereby avoiding the introduction of variational inequalities. In this case, only gradient equations need be considered, and the methods of convex analysis, coupled to the Schauder theorem, suffice. An interesting approach to the system, augmented by a coupling induced by laser beam modeling, is furnished in [21].

4.1 Contraction Property of the Gummel Map in Two Dimensions

We consider the steady-state drift-diffusion model, subject to the following simplifying conventions. We permit the device to include an insulation buffer $\bar{G}_b$, separating the gate from the substrate, $\bar{G}$.

1. The fundamental dependent variables are u and the modified Slotboom variables, ν, ω, so that electron and hole carrier concentrations are given, respectively, by
$$n = e^{u-u_*}\nu, \qquad p = e^{-(u-u_*)}\omega, \tag{4.4}$$
where u_* is the intrinsic semiconductor potential. This may be thought of as an equilibrium potential, and is inserted because of the estimate (4.22) below. The potential u is defined on the entire device, consisting of buffer and substrate, whereas ν and ω are defined on the substrate, $\bar{G}$.
2. Zero recombination, constant carrier mobility and diffusion, and the Einstein relations are assumed, at ambient temperature, T_0.
3. The charge units are volts, and the potential (and, implicitly, the associated Fermi levels) are dimensionless. By band-bending in this section is meant the maximal variation of the potential, u.
4. Boundary values for the potential are specified on the contacts; the base is grounded. The doping levels define n and p on source, drain, and base, so that ν and ω can be determined from (4.4). Insulation-type conditions prevail elsewhere, including the curve between substrate and buffer, where the normal component of the current is zero; the products of the normal field components with the dielectric constants must also be equal along this curve to ensure normal continuity of the electric displacement vector. These conditions are understood as augmenting the equations below. The device is assumed to be a two-dimensional MOS-FET device, so that a schematic representation of buffer rectangle symmetrically atop a larger substrate rectangle is adequate.

The drift-diffusion model thus assumes the following form in steady-state. The potential u is harmonic on G_b, and,

$$\begin{aligned} -\epsilon\nabla^2 u + e^{u-u_*}\nu - e^{-(u-u_*)}\omega &= k_1, \quad \text{on } G, & (4.5)\\ -\nabla\cdot(e^{u-u_*}\nabla\nu) &= 0, \quad \text{on } G, & (4.6)\\ -\nabla\cdot(e^{u-u_*}\nabla\omega) &= 0, \quad \text{on } G, & (4.7)\end{aligned}$$

subject to boundary values for u, ν, and ω. Here k_1 describes the doping in the device, and ϵ is the dielectric constant in the substrate, G.

4.1.1 A Framework for the Gummel Map

The Gummel map $\mathbf{T}_s$ may be viewed as a solution map associated with (4.5–4.7). The subscript emphasizes the role played by the Slotboom variables; distinguishing subscripts are used later in the chapter when quasi-Fermi levels are employed. The definition adopted in this section is as follows.

Definition 4.1.1. *Given a pair* $[\bar{\nu}, \bar{\omega}]$ *satisfying the inequalities,*

$$0 < \alpha_\nu = \inf \bar{\nu} \leq \bar{\nu} \leq \sup \bar{\nu} = \beta_\nu, \tag{4.8}$$

$$0 < \alpha_\omega = \inf \bar{\omega} \leq \bar{\omega} \leq \sup \bar{\omega} = \beta_\omega, \tag{4.9}$$

where $\bar{\nu}$ *and* $\bar{\omega}$ *denote boundary values of* ν *and* ω *on source, gate, drain, and base, we specify* $\tilde{u} = \tilde{u}(\bar{\nu}, \bar{\omega})$ *in terms of the fractional step: solve* (4.5) *with* $\nu = \bar{\nu}$ *and* $\omega = \bar{\omega}$, *subject to boundary values for* u. *Given* $\tilde{u}$, *set*

$$[\nu, \omega] := \mathbf{T}_s(\bar{\nu}, \bar{\omega}), \tag{4.10}$$

where ν *and* ω *solve the linear, uncoupled equations on* G *given by*

$$-\nabla \cdot (e^{\tilde{u}-u_*} \nabla \nu) = 0, \tag{4.11}$$

$$-\nabla \cdot (e^{-(\tilde{u}-u_*)} \nabla \omega) = 0, \tag{4.12}$$

subject to the specified boundary-values for ν *and* ω.

If K denotes the closed, convex set of square-integrable pairs satisfying (4.8, 4.9), it can be shown that the following properties hold.

Property 4.1.1. a) K is invariant under $\mathbf{T}_s$ (maximum principles).

b) $\mathbf{T}_s$ has a fixed point, $[\nu, \omega]$, and $(\tilde{u}(\nu, \omega), \nu, \omega)$ is a solution of the drift-diffusion system.

c) The solutions of (4.11, 4.12), as well as the defining equation for $\tilde{u}$, are understood in the weak sense; still, the gradients of u, ν, and ω are square integrable, i.e. , are of finite energy. On the basis of known results (cf. [42]), we shall assume that ν and ω satisfy estimates of the form,

$$|\nabla \nu(\mathbf{x})| \leq c r^{-1/2}, \quad |\nabla \omega(\mathbf{x})| \leq c r^{-1/2}, \tag{4.13}$$

for some positive constant c independent of $\bar{\nu}$ and $\bar{\omega}$; here r denotes the minimal distance to a finite number of singular boundary points. We do not *exclude* singularities for u, but they do not enter explicitly here. Such singularities in $\nabla \nu$ and $\nabla \omega$ can occur because of transition points between Dirichlet and Neumann boundary data.

These facts are documented later in this chapter. There, the existence statement is proven by the (nonconstructive) Schauder fixed point theorem, which can hold even if the Picard iterates fail to converge. However, the convergence of these iterates is highly desirable, and is guaranteed if

$$\int_G \{|\nu_1 - \nu_2|^2 + |\omega_1 - \omega_2|^2\}\, dx \le C^2 \int_G \{|\tilde{\nu}_1 - \tilde{\nu}_2|^2 + |\tilde{\omega}_1 - \tilde{\omega}_2|^2\}\, dx, \quad (4.14)$$

for $C < 1$, where $[\nu_i, \omega_i] = \mathbf{T}_s(\tilde{\nu}_i, \tilde{\omega}_i)$ for $i = 1, 2$. The theorem of this section, presented in the next subsection and proved in the following one, displays C explicitly in terms of the parameters of the device, and provides estimates for when $C < 1$ is to be expected. C is called a Lipschitz constant for $\mathbf{T}_s$. In particular, if $C < 1$, one can choose $[\nu_0, \omega_0]$ arbitrarily in K, and the Picard iterates,

$$(\nu_m, \omega_m) = \mathbf{T}_s^m(\nu_0, \omega_0), \quad (4.15)$$

will converge to the unique fixed point of $\mathbf{T}_s$ in K as $m \to \infty$. This is also a situation when the device equations have a unique solution.

4.1.2 The Principal Result

We proceed to the major result of the section. Following the statement of Theorem 4.1.1 are some elementary numerical estimates of the Lipschitz constant. The four supporting lemmas are stated and proved in the following subsection, with their hypotheses. The proof of Theorem 4.1.1 follows directly from concatenation of Lemmas 4.1.1–4.1.4.

Theorem 4.1.1. *Suppose hypotheses* (Ha)–(Hd) *of* §4.1.3 *hold, and that the boundary data, $\bar{u}$, $\bar{\nu}$, and $\bar{\omega}$ are pointwise bounded restrictions of finite energy functions, with $\alpha_\nu > 0$, $\alpha_\omega > 0$ in* (4.8, 4.9), *and that the doping k_1 is pointwise bounded. Then the Gummel map $\mathbf{T}_s$ is well defined by* (4.10), *and the closed, convex subset K of pairwise square-integrable functions $[\tilde{\nu}, \tilde{\omega}]$, defined by* (4.8, 4.9), *is invariant under $\mathbf{T}_s$. On K, the map $\mathbf{T}_s$ satisfies* (4.14), *with C explicitly estimated by*

$$C = 2\sqrt{2}\, c\epsilon^{-1} \left[d^5 \left(\ln\left(\frac{d}{d_0}\right) + 1 \right) \right]^{1/2} \cdot e^{\delta-\gamma} \max(e^{\delta - u_*}, e^{-(\gamma - u_*)}). \quad (4.16)$$

The constants c, δ, and γ are described in (4.13) *and* (4.20, 4.21), *the dielectric constant ϵ is introduced in* (4.5), *and u_* is the intrinsic potential; d_0 is defined in* (4.35) *below.*

Remark 4.1.1. If d and d_0 are of comparable magnitude, and this relation is maintained as d decreases, then

$$C = O(d^{5/2}), \qquad d \to 0. \quad (4.17)$$

This theorem characterizes the Gummel solution map for device simulation, as based on the Slotboom variables, restricted pointwise to range between minimum and maximum boundary values, as guaranteed by the maximum principles. Zero recombination, constant mobilities, and Einstein relations have been assumed. The Lipschitz constant C of the map has been estimated via inequality (4.16). In particular, if $C < 1$, then the successive approximations, defined by the Picard iterates, (4.15), converge to the pair $[\nu, \omega]$, for which $(\tilde{u}(\nu,\omega), \nu, \omega)$ is the unique solution of the device system (4.5–4.7), subject to the prescribed contact boundary values. Here, $\tilde{u}(\nu,\omega)$ denotes the solution of (4.5). According to well-known estimates,

$$\|[\nu,\omega] - [\nu_m,\omega_m]\| \leq \frac{C^m}{1-C}\|[\nu_1,\omega_1] - [\nu_0,\omega_0]\| \tag{4.18}$$

if $C < 1$, where the norm-square difference is defined by the left hand side of (4.14).

The significance of this analysis is the derivation of the asymptotic dependence of C on the device (actually substrate) diameter d, and the numerical estimation of C itself. If units are employed as in [37] (cf. §2.3.1), then for a micron diameter device, for which $d = 1$,

$$2\sqrt{2}\,\epsilon^{-1}c\left[d^5\left(\ln\left(\frac{d}{d_0}\right)+1\right)\right]^{1/2} \leq 0.0355c, \tag{4.19}$$

if $d/d_0 \leq 2$, and $\epsilon = 11.7 \times 8.85$ is chosen as the dielectric constant for silicon in these units. An analysis of the remaining factors in (4.16) requires the inequalities,

$$u \geq \gamma = \min\left[\inf \bar{u}, \ln\left\{\frac{\inf k_1 + \sqrt{(\inf k_1)^2 + 4\beta_\nu\alpha_\omega}}{2\beta_\nu}\right\}\right], \tag{4.20}$$

$$u \leq \delta = \max\left[\sup \bar{u}, \ln\left\{\frac{\sup k_1 + \sqrt{(\sup k_1)^2 + 4\alpha_\nu\beta_\omega}}{2\alpha_\nu}\right\}\right]. \tag{4.21}$$

Here, $\bar{u}$ is defined on Σ_D, i.e. , on source, gate, drain, and base, and k_1 on the substrate, $\bar{G}$. The inf-sup operations are taken over the domains of definition. The analysis then shows that, for C not to exceed 1, about $3.337 - \ln c$ constitutes the maximum allowable spread for

$$\Delta u + \Delta u_i = [(\delta - \gamma) + (\delta - u_*)], \tag{4.22}$$

in the case, say, of electron majority carriers. Here Δu is the maximum band-bending, and Δu_i is the maximum shift of the intrinsic potential level. Heuristic methods of estimating c are given in [68].

Although the estimate above, for C not to exceed unity, may seem unduly restrictive, it is consistent with present computing strategies, loosely bundled into the category of continuation on the bias potential, which is related to the band bending. Experience has shown that convergence of iterative methods is to be expected when this continuation is effected with sufficiently small incremental bias steps, and the analysis confirms this. Results related to those described in the theorem above, in that, what we interpret here as band bending, is critically analyzed, have been obtained in [104] and [83].

The latter contains an interesting result on the approach of iterates far from the solution, and an estimate of when this phenomenon ceases, in reference to the metric distance from the solution.

We can obtain uniqueness only for $C < 1$, as a consequence of the contraction mapping principle. In this case, the mapping is a strict contraction. The reader interested in specific studies of uniqueness in one dimension is invited to consult [2], which contains an informative literature survey. In particular, a discussion of nonuniqueness is given in this reference, and the relation of this question to the doping oscillation is pinpointed, especially with respect to the thyristor, a device with three junctions. A study of multiple states in the thyristor, in the case of the local electroneutrality hypothesis, has been carried out in [114]. One can build on these results, via the analysis of [86], to cover the case of nonzero space charge approximation. It also follows from the results of [2] that, when the boundary conditions are such that the electric field must be of one sign, then the model without doping must have a unique solution. In a later publication [3], the case of (skew-) symmetric solutions is considered. Finally, the result of [20], proven for a sufficiently small parameter proportional to a power of the device length, may be viewed as a one-dimensional version of our contraction principle stated above.

4.1.3 The Supporting Lemmas and Hypotheses

The first lemma is a generalized Poincaré inequality , comparing the integral mean square values of a function and its gradient, provided the function vanishes on a significantly extensive subset Σ of the boundary ∂G of G. It appears to be a new result; we have not encountered it in the literature. For this lemma, we permit $G \subset \mathbb{R}^n$, and we require two hypotheses.

Assumption 4.1 (first two hypotheses). (Ha) It is possible to write $\bar{G}$ as the union of a finite number of disjoint subsets K_j, such that K_j is covered by parallel segments, emanating from $K_j \cap \Sigma$, aligned with one of the coordinate axes. The choice of the coordinate axis may change with j. Here Σ is a specified subset of ∂G.

(Hb) It is possible to write the multiple integral of a smooth function over each K_j in (Ha) as an iterated integral, in either order, where the measures are dx_i and $dx_1 \cdots dx_{i-1} dx_{i+1} \cdots dx_n$, respectively. Here. $i = i(j)$ is specified according to that coordinate axis parallel to the covering lines.

Lemma 4.1.1. *Let* $u \in C^\infty(\bar{G})$ *with* $u|_\Sigma = 0$. *Then, if hypotheses* (Ha) *and* (Hb) *hold, the inequality,*

$$\int_G |u(\mathbf{x})|^2\, dx \le d^2 \int_G |\nabla u(\mathbf{x})|^2\, dx, \tag{4.23}$$

is valid, where d = *diameter* $(G) = \sup\{|\mathbf{P} - \mathbf{Q}| : \mathbf{P}, \mathbf{Q} \in G\}$. *By completion, inequality* (4.23) *holds for finite energy functions with zero trace on* Σ.

Proof. Assume, for simplicity, that $G = K_1$, and that the coordinate axis chosen is the x_1 axis. Then, for each $\mathbf{x} \in G$, there is guaranteed, by hypothesis (Ha), an $\mathbf{x}_0 = \mathbf{x}_0(\mathbf{x}) \in G$ such that

$$u(\mathbf{x}) = \int_{\mathbf{x}_0}^{\mathbf{x}} \frac{\partial u}{\partial x_1}(x_1, x_2, \ldots, x_n)\, dx_1, \tag{4.24}$$

and, by the Schwarz inequality,

$$|u(\mathbf{x})|^2 \le d \int_{S(\mathbf{x}_0)} \left| \frac{\partial u}{\partial x_1}(x_1, x_2, \ldots, x_n) \right|^2 dx_1, \tag{4.25}$$

where $S(\mathbf{x}_0)$ is the entire line segment, emanating from or terminating at $\mathbf{x}_0$, contained in G, and parallel to the x_1 axis. By the assumption (Hb), on the iteration of multiple integration, and, by (4.25), there is an interval (a, b) such that

$$\begin{aligned}
\int_G |u(\mathbf{x})|^2\, dx &= \int_a^b \left(\int_{\pi(x_1)} \cdots \int |u(\mathbf{x})|^2\, dx_2 \cdots dx_n \right) dx_1 \\
&\le \int_a^b \left(d \int_{\pi(x_1)} \cdots \int \left(\int_{S(\mathbf{x}_0)} \left(\frac{\partial u}{\partial x_1} \right)^2 dx_1 \right) dx_2 \cdots dx_n \right) dx_1
\end{aligned}$$

where $\pi(x_1) \subset G$ is orthogonal to $S(\mathbf{x}_0)$ for each x_1. Since

$$\int_{\pi(x_1)} \left(\int_{S(\mathbf{x}_0)} \left(\frac{\partial u}{\partial x_1} \right)^2 dx_1 \right) dx_2 \cdots dx_n \le \int_G |\nabla(u(\mathbf{x}))|^2\, dx$$

for each $a \le x_1 \le b$, it follows that

$$\int_G |u(\mathbf{x})|^2\, dx \le (b-a) d \int_G |\nabla(u(\mathbf{x}))|^2\, dx,$$

which implies (4.23), since $(b - a) \le d$. If there are several regions K_j, their individual contributions are computed as above and added.

Remark 4.1.2. This result will be applied to the differences on the left hand side of (4.14), corresponding to the solutions associated with different pairs $[\tilde{\nu}_1, \tilde{\omega}_1]$ and $[\tilde{\nu}_2, \tilde{\omega}_2]$ in K. The base of the device alone, taken as Σ, is sufficiently extensive to apply Lemma 4.1.1 to such differences.

The next lemma relates how the gradient differences can be bounded in terms of $\tilde{u}_1 - \tilde{u}_2$, as measured by an appropriate weight, W. We require the following hypothesis (cf. (4.13)), which is consistent with known results (cf. [7, 137]).

Assumption 4.2 (third hypothesis). (Hc) For any pair $[\tilde{\nu}, \tilde{\omega}]$ in K, the Gummel map produces a pair $[\nu, \omega] = \mathbf{T}_s(\tilde{\nu}, \tilde{\omega})$, with gradient singularity of order $r^{-1/2}$. More precisely, there are finitely many points $\mathbf{P}_1, \cdots, \mathbf{P}_\ell$ on ∂G, and a positive constant c, such that

$$|\nabla\nu(\mathbf{x})|^2 \le W(\mathbf{x}), \quad |\nabla\omega(\mathbf{x})|^2 \le W(\mathbf{x}), \qquad \mathbf{x} \in G, \tag{4.26}$$

where W is the weight function

$$W(\mathbf{x}) = c^2 \max_{1\le i\le \ell} [\text{distance}(\mathbf{x}, \mathbf{P}_i)]^{-1}, \qquad \mathbf{x} \in G. \tag{4.27}$$

Weaker singularities are, of course, included.

Lemma 4.1.2. *Suppose the hypothesis* (Hc) *holds. Set*

$$c_1 = 2e^{2(\delta-\gamma)}. \tag{4.28}$$

Then the inequality,

$$\int_G \{|\nabla(\nu_1 - \nu_2)|^2 + |\nabla(\omega_1 - \omega_2)|^2\}\, dx \le c_1 \int_G |\tilde{u}_1 - \tilde{u}_2|^2 W\, dx, \tag{4.29}$$

holds. Here, $\tilde{u}_i = \tilde{u}_i(\tilde{\nu}_i, \tilde{\omega}_i)$, *and* $[\nu_i, \omega_i] = \mathbf{T}_s(\tilde{\nu}_i, \tilde{\omega}_i)$, $i = 1, 2$.

Proof. We start with the weak form of (4.11):

$$\int_G e^{\tilde{u}-u_*} \nabla\nu \cdot \nabla\phi\, dx = 0, \tag{4.30}$$

for all finite energy functions ϕ, with zero trace on substrate contacts. If (4.30) is employed with $\tilde{u} = \tilde{u}_1$ and $\tilde{u} = \tilde{u}_2$, and the relations are subtracted, one obtains

$$\int_G e^{\tilde{u}_1-u_*} \nabla(\nu_1 - \nu_2) \cdot \nabla\phi\, dx = -\int_G [e^{\tilde{u}_1-u_*} - e^{\tilde{u}_2-u_*}] \nabla\nu_2 \cdot \nabla\phi\, dx.$$

Setting $\phi = \nu_1 - \nu_2$, and using the inequalities,

$$e^{\tilde{u}_1-u_*} \ge e^{\gamma-u_*}, \qquad ab \le \frac{1}{2}(a^2 + b^2),$$

we obtain

$$\begin{aligned} \int_G |\nabla(\nu_1 - \nu_2)|^2\, dx &\le e^{-2(\gamma-u_*)} \int_G |e^{\tilde{u}_1-u_*} - e^{\tilde{u}_2-u_*}|^2\, |\nabla\nu_2|^2\, dx \\ &\le e^{2(\delta-\gamma)} \int_G |\tilde{u}_1 - \tilde{u}_2|^2\, W\, dx, \end{aligned} \tag{4.31}$$

where we have used the mean-value theorem and the inequalities,

$$e^{\tilde{u}_i-u_*} \le e^{\delta-u_*}, \qquad i = 1, 2.$$

Similar arguments show that

$$\int_G |\nabla(\omega_1 - \omega_2)|^2\, dx \le e^{2(\delta-\gamma)} \int_G |\tilde{u}_1 - \tilde{u}_2|^2\, W\, dx, \tag{4.32}$$

so that (4.29) follows from (4.31) and (4.32).

Remark 4.1.3. The next lemma is a technical extension of Lemma 4.1.1, when the integration of u involves the weight W. The intent is to estimate the right hand side of (4.29) in terms of gradients.

We require the following hypothesis, permitting polar coordinate representations.

Assumption 4.3 (final hypothesis). (Hd) The substrate $\bar{G}$ may be decomposed into disjoint subsets K_j, $1 \le j \le \ell$, such that $\mathbf{P}_j \in K_j$, $|\mathbf{x} - \mathbf{P}_j| = \min_{1 \le i \le \ell} |\mathbf{x} - \mathbf{P}_i| (\mathbf{x} \in K_j)$, and $\mathbf{P}_j$ is at a positive distance from $K_j \cap \Sigma_D^*$, where $\mathbf{P}_j$ is any one of the singular points of (Hc), and Σ_D^* is a subset of the Dirichlet boundary, with the following property. For each $u \in C^\infty(\bar{G})$, vanishing on $\Sigma_D^* \cap \bar{G}$, there is a line integral representation in polar coordinates, with $\mathbf{P} = \mathbf{P}_j$ as *origin*, given by

$$u(\mathbf{x}) = \int_{r'=r(\mathbf{x}_0)}^{r'=r(\mathbf{x})} \frac{\partial u}{\partial r'}\, dr' + \int_{\theta'=\theta(\mathbf{x}_0)}^{\theta'=\theta(\mathbf{x})} \frac{\partial u}{\partial \theta'}\, d\theta', \tag{4.33}$$

where $\mathbf{x}_0 = \mathbf{x}_0(\mathbf{x}) \in K_j \cap \Sigma_D^*$. In (4.33), $\theta = \theta(\mathbf{x}_0)$ in the first integral, and $r = r(\mathbf{x})$ in the second integral.

Lemma 4.1.3. *Let $d_j =$ distance $(\mathbf{P}_j, K_j \cap \Sigma_D^*)$. Then, for u as in* (Hd) *and $1 \le j \le \ell$,*

$$\int_{K_j} |\mathbf{x} - \mathbf{P}_j|^{-1} |u(\mathbf{x})|^2\, dx \le 2d \left[\ln\left(\frac{d}{d_j}\right) + 1\right] \int_{K_j} |\nabla(u(\mathbf{x}))|^2\, dx. \tag{4.34}$$

If

$$d_0 = \min_{1 \le j \le l} d_j, \tag{4.35}$$

then

$$\int_G |u(\mathbf{x})|^2\, W(\mathbf{x})\, dx \le 2c^2 d \left[\ln\left(\frac{d}{d_0}\right) + 1\right] \int_G |\nabla u(\mathbf{x})|^2\, dx, \tag{4.36}$$

where c is prescribed in (4.27).

Proof. We use (4.33) as the starting point, together with the relations,

$$\begin{aligned} \frac{\partial u}{\partial r} &= \nabla u \cdot (\cos\theta, \sin\theta), \\ \frac{\partial u}{\partial \theta} &= \nabla u \cdot r(-\sin\theta, \cos\theta). \end{aligned} \tag{4.37}$$

With the help of the Schwarz inequality, and the inequality $(a+b)^2 \le 2a^2 + 2b^2$, we obtain, with $\tilde{d}_j$ the maximal distance to the specified boundary segment,

$$\begin{aligned} |u(\mathbf{x})|^2 &\leq 2\left(\int_{r'=0}^{r'=r^*(\mathbf{x}_0)} |\nabla u|^2 r'\,dr'\right) \cdot \left|\int_{r'=\tilde{d}_j}^{r'=r(\mathbf{x})} (r')^{-1}\,dr'\right| \\ &\quad + 2|r(\mathbf{x})|^2 \int_{S(\mathbf{x})} |\nabla u|^2\,d\theta' \end{aligned} \tag{4.38}$$

where $r^*(\mathbf{x}_0)$ is the maximal extent in G of a ray, located by the angle, $\theta(\mathbf{x}_0)$, and $S(\mathbf{x})$ is the total range of θ' in G for $r = r(\mathbf{x})$. Inequality (4.38) may be rewritten as

$$\begin{aligned} |u(\mathbf{x})|^2 &\leq 2\left|\ln\left(\frac{r(\mathbf{x})}{\tilde{d}_j}\right)\right| \int_{r'=0}^{r'=r^*(\mathbf{x}_0)} |\nabla u|^2\, r'\,dr' \\ &\quad + 2|r(\mathbf{x})|^2 \int_{S(\mathbf{x})} |\nabla u|^2\,d\theta'. \end{aligned} \tag{4.39}$$

Integrating the product of (4.39), with the weight $|\mathbf{x} - \mathbf{P}_j|^{-1} = r^{-1}$, over K_j, and using the inequalities, $r \leq d$, $d_j \leq \tilde{d}_j$, we have

$$\begin{aligned} \int_{K_j} r^{-1}|u|^2 r\,dr\,d\theta &\leq 2\int_{r=0}^{r=d} \left|\ln\left(\frac{r}{\tilde{d}_j}\right)\right| dr \int_{K_j} |\nabla u|^2\, r'\,dr'\,d\theta \\ &+ 2d\int_{K_j} |\nabla u|^2 r\,dr\,d\theta' \\ &\leq 2d\left[\ln\left(\frac{d}{d_j}\right) - 1\right] \int_{K_j} |\nabla u|^2\, r'\,dr'\,d\theta \\ &+ 2d\int_{K_j} |\nabla u|^2 r\,dr\,d\theta' \\ &= 2d\left[\ln\left(\frac{d}{d_j}\right) + 1\right] \int_{K_j} |\nabla u|^2\,dx. \end{aligned}$$

We have thus derived inequality (4.34). Inequality (4.36) is immediate.

Remark 4.1.4. The final lemma estimates the gradient norm of $\tilde{u}_1 - \tilde{u}_2$, in terms of $\tilde{\nu}_1 - \tilde{\nu}_2$ and $\tilde{\omega}_1 - \tilde{\omega}_2$.

Lemma 4.1.4. *Let $\bar{G}_b$ denote the buffer. Then*

$$\int_{G\cup G_b} |\nabla(\tilde{u}_1 - \tilde{u}_2)|^2\,dx \leq c_2 d^2 \left[\int_G |\tilde{\nu}_1 - \tilde{\nu}_2|^2\,dx + \int_G |\tilde{\omega}_1 - \tilde{\omega}_2|^2\,dx\right] \tag{4.40}$$

for any $[\tilde{\nu}_1, \tilde{\omega}_1]$, $[\tilde{\nu}_2, \tilde{\omega}_2]$ in K, where c_2 is given by

$$c_2 = 2\epsilon^{-2}\max(e^{2(\delta - u_*)}, e^{-2(\gamma - u_*)}). \tag{4.41}$$

Proof. We begin by noting the weak form of (4.5) (where ϵ_0 is the dielectric constant associated with G_b):

$$\epsilon_0 \int_{G_b} \nabla u \cdot \nabla \phi \, dx + \epsilon \int_G \nabla u \cdot \nabla \phi \, dx + \int_G (e^{u-u_*}\nu - e^{-(u-u_*)}\omega - k_1)\phi \, dx = 0, \tag{4.42}$$

for all finite energy functions ϕ vanishing on all of the device contacts. Given $\nu = \tilde{\nu}_i$, $\omega = \tilde{\omega}_i$, we solve (4.42) for $\tilde{u}_i$, $i = 1, 2$, subtract the two relations, and set $\phi = \tilde{u}_1 - \tilde{u}_2$ to obtain, after neglecting the buffer contribution,

$$\begin{aligned}\epsilon \int_G |\nabla(\tilde{u}_1 - \tilde{u}_2)|^2 \, dx \quad &+ \int_G e^{\tilde{u}_2 - u_*}(e^{\tilde{u}_1 - \tilde{u}_2} - 1)\tilde{\nu}_1(\tilde{u}_1 - \tilde{u}_2) \, dx \\ &+ \int_G e^{-(\tilde{u}_2 - u_*)}(1 - e^{-(\tilde{u}_1 - \tilde{u}_2)})\tilde{\omega}_1(\tilde{u}_1 - \tilde{u}_2) \, dx \\ &\le - \int_G e^{\tilde{u}_2 - u_*}(\tilde{\nu}_1 - \tilde{\nu}_2)(\tilde{u}_1 - \tilde{u}_2) \, dx \\ &+ \int_G e^{-(\tilde{u}_2 - u_*)}(\tilde{\omega}_1 - \tilde{\omega}_2)(\tilde{u}_1 - \tilde{u}_2) \, dx. \end{aligned} \tag{4.43}$$

The right hand side of (4.43) is bounded by

$$\frac{1}{2}\epsilon d^{-2} \int_G |\tilde{u}_1 - \tilde{u}_2)|^2 \, dx \; +$$

$$\epsilon^{-1} d^2 \cdot \max(e^{2(\delta - u_*)}, e^{-2(\gamma - u_*)}) \cdot \int_G \{|\tilde{\nu}_1 - \tilde{\nu}_2|^2 \; + |\tilde{\omega}_1 - \tilde{\omega}_2|^2\} \, dx.$$

Moreover, the second and third terms on the left hand side are non-negative. After absorption, we conclude from (4.43), from (4.23), and from the above bound that

$$\int_G |\nabla(\tilde{u}_1 - \tilde{u}_2)|^2 \, dx \le$$

$$2\epsilon^{-2} d^2 \max(e^{2(\delta - u_*)}, e^{-2(\gamma - u_*)}) \cdot \left[\int \{|\tilde{\nu}_1 - \tilde{\nu}_2|^2 \; + |\tilde{\omega}_1 - \tilde{\omega}_2|^2\} \, dx\right],$$

which leads to inequality (4.40), when the buffer term is included.

Remark 4.1.5 (proof summary). By Lemma 4.1.1, the left hand side of (4.14) is bounded by (with $\Sigma :=$ base, $n = 2$, and $K_1 = G$)

$$d^2 \int_G \{|\nabla(\nu_1 - \nu_2)|^2 + |\nabla(\omega_1 - \omega_2)|^2\} \, dx,$$

and the latter expression, by Lemma 4.1.2, is bounded by

$$c_1 d^2 \int_G |\tilde{u}_1 - \tilde{u}_2|^2 \, W \, dx, \qquad c_1 := 2e^{2(\delta - \gamma)}.$$

By Lemma 4.1.3, the previous bound translates to

$$2c^2 c_1 d^3 \left[\ln\left(\frac{d}{d_0}\right) + 1\right] \int_G |\nabla(\tilde{u}_1 - \tilde{u}_2)|^2 \, dx.$$

An application of Lemma 4.1.4 now yields estimate (4.16) of Theorem 4.1.1.

4.2 General Case: A Framework of Weighted Spaces

We begin this section by stating the major hypothesis concerning transition point singularities which is made for the remainder of this chapter. This hypothesis, a weakening of Assumption 4.2, is made in terms of a lower bound of polynomial decay for the mobility coefficients near the transition points. The decay is assumed of uniform order when evaluated at arbitrary electric field values, arising from the range of the map $\mathbf{U}_f$. This map has already been introduced in other contexts, but a careful definition in terms of weak solutions and test functions is provided for reference in the next section. We note here that the domain of $\mathbf{U}_f$ is given by (4.51). Although this hypothesis is inherently designed to cover the case $N \geq 2$, the case $N = 1$ can be retrieved by the selection of $\theta = 0$ in what follows. This choice also includes mobilities which are only position dependent.

Assumption 4.4 (polynomial decay). Let $\mathcal{P}$ denote the transition boundary points on ∂G. It is assumed that there are real numbers θ and C, $0 \leq \theta < N$, $C > 0$, such that, for any field value, $\mathbf{E} = -\nabla u$, where u is in the range of $\mathbf{U}_f$, the lower bound estimates for the decay,

$$\mu_n(\mathbf{E}(\mathbf{x})) \geq C \inf_{\mathbf{P}\in\mathcal{P}} |\mathbf{P} - \mathbf{x}|^{\theta}, \tag{4.44}$$

$$\mu_p(\mathbf{E}(\mathbf{x})) \geq C \inf_{\mathbf{P}\in\mathcal{P}} |\mathbf{P} - \mathbf{x}|^{\theta}, \tag{4.45}$$

hold for almost all $\mathbf{x} \in G$.

In order to place this assumption in perspective, we note that it is implied, for mobilities of the form (4.47) below, by the following.

Assumption 4.5 (electric field growth). It is assumed that there are real numbers θ and c, $0 \leq \theta < N$, $c > 0$, such that the following pointwise almost everywhere gradient estimate holds on G:

$$|E(\mathbf{x})| \leq c \sup_{\mathbf{P}\in\mathcal{P}} |\mathbf{P} - \mathbf{x}|^{-\theta}. \tag{4.46}$$

Here, $E = -\nabla u$ for arbitrary u in the range of $\mathbf{U}_f$.

Remark 4.2.1. Thus, for arbitrary $\kappa \geq 1$, and for mobilities given by

$$\mu = \frac{c_1}{(1 + c_2|\mathbf{E}|^{\kappa})^{1/\kappa}}, \tag{4.47}$$

it follows that (4.46) implies (4.44) and (4.45), with the same exponential index, θ. The mobility relations (4.47) were introduced and discussed in chapter one (cf. (2.42)). Although $\mathbf{U}_f$ is technically a nonlinear Poisson solver for fixed boundary values, it is very simple to reformulate the preceding hypothesis in terms of an equivalent linear solver. For $[v, w]$ given, set

$$n = \exp(\mathbf{U}_f[v, w] - v), \quad p = \exp(w - \mathbf{U}_f[v, w]),$$

and solve the corresponding *linear* Poisson equation for $u = \mathbf{U}_f[v, w]$, as specified by uniqueness. The advantage of an assumption stated in terms of the linear Poisson equation, and the corresponding asymptotic rate (4.46), is that a substantial literature already exists, at least in two dimensions ([137], [7]). In these studies, and in [42], a value of θ of $\frac{1}{2}$ is predicted for transition points on a linear segment; in [42], a careful study is made for a transition point adjacent to the silicon interface of a MOS-FET device, and an analytical result is presented, involving relative dielectrics. This result suggests that the source and drain contacts should have their endpoints away from the oxide interface. It is important to mention here that the results obtained in [42] are systems' results, and the diffusion coefficients for the current continuity subsystem were assumed bounded away from zero.

These hypotheses permit us to define a class of weighted Sobolev spaces. We describe these now. Denote by ρ the common right hand sides of inequalities (4.44) and (4.45):

$$\rho(\mathbf{x}) = C \inf_{\mathbf{P} \in \mathcal{P}} |\mathbf{P} - \mathbf{x}|^{\theta}. \tag{4.48}$$

This spatially dependent function will serve as the appropriate function space weight in the weighted space, $H^1_\rho(G)$, with nonstandard norm given by

$$\|f\|^2_{H^1_\rho} = \int_G \rho |\nabla f|^2 + \left(\int_{\Sigma_D} \Gamma f\right)^2. \tag{4.49}$$

The properties of these weighted Sobolev spaces have been studied in [92], where they are characterized as spaces of locally L_2 functions with distribution gradients, such that the first term on the right hand side of (4.49) is finite. The definition given by (4.48) is a slight generalization of [92], where the case of only a finite number of points is considered. The norm employed in [92] is not the norm (4.49), but rather the obvious extension of the H^1 norm, so that ρ appears as a weight in each integral:

$$\|f\|^2 = \int_G \rho(|f|^2 + |\nabla f|^2). \tag{4.50}$$

We relegate to the final section the technical verification that (4.50) and (4.49) are equivalent norms. We shall employ (4.49) throughout. In [52], certain weighted space results are derived, motivated by geometric singularities, but we shall not use those results here. We have not explicitly stated hypotheses on the regularity of ∂G, but it must be sufficient for standard trace inequalities (cf. (4.115)) and continuous and compact embeddings (cf. (4.114) and (4.92)) to hold.

4.3 Existence and Maximum Principles for $\mathbf{U}_f$

We begin by discussing the definition of the mapping $\mathbf{U}_f$. Specifically, we restrict the domain to be the closed, convex set K in $L_2(G) \times L_2(G)$, where

$$K = \{[v, w] : \alpha \leq v \leq \beta, \alpha \leq w \leq \beta\}, \tag{4.51}$$

with inequalities taken almost everywhere, and where

$$\alpha = \min(\inf_{\Sigma_D} \bar{v}, \inf_{\Sigma_D} \bar{w}), \tag{4.52}$$

$$\beta = \max(\sup_{\Sigma_D} \bar{v}, \sup_{\Sigma_D} \bar{w}). \tag{4.53}$$

The precise definition of $\mathbf{U}_f$, which is used as a fractional step in the definition of the fixed point map $\mathbf{T}_f$, entails the solution of (4.54) below, subject to the specified boundary values given by (4.55).

Lemma 4.3.1. *Given a pair $[v, w]$ in K, the image under the map $\mathbf{U}_f$ is the unique element u satisfying the weak relation* (cf. (4.1)),

$$\langle \epsilon \nabla u, \nabla \phi \rangle + \langle \exp(u - v) - \exp(w - u) - k_1, \phi \rangle, \tag{4.54}$$

subject to the boundary condition,

$$\Gamma(u - \bar{u}) \,|_{\Sigma_D} = 0. \tag{4.55}$$

The test function space in this relation, denoted $Y_0 := H^1_{0,\Sigma_D}$, consists of H^1 functions with zero trace on Σ_D. Moreover, u satisfies the maximum principle,

$$u \geq \gamma = \min(\inf_{\Sigma_D} \bar{u}, \gamma'), \tag{4.56}$$

$$u \leq \delta = \max(\sup_{\Sigma_D} \bar{u}, \delta'), \tag{4.57}$$

where γ' and δ' are uniquely defined by

$$2 \sinh(\gamma' - \alpha) - \inf_G k_1 = 0, \tag{4.58}$$

$$2 \sinh(\delta' - \beta) - \sup_G k_1 = 0. \tag{4.59}$$

Proof. We first demonstrate the validity of the bounds (4.56, 4.57). Thus, if u is a solution in H^1, with $\exp u$ in L_2, we select for the admissible test function ϕ, $\phi = (u - \delta)^+$, where $t^+ = \max\{t, 0\}$. The restriction of the trace of ϕ to Σ_D is zero, hence ϕ is admissible, since the positive part is subadditive, and this property may be applied to $u - \delta = U_0 + (\bar{u} - \delta)$, for some admissible U_0. Now the substitution of ϕ into (4.54) reduces integrations to the set $\{u > \delta\}$. For the term involving the gradients, this uses the chain rule for the composition of the positive part and an H^1 function ([48, Lemma 7.6]). Once this reduction is achieved, one uses the nonnegativity of

$$\exp(u - v) - \exp(w - u) - k_1,$$

on the set $\{u > \delta\}$, to conclude $\nabla(u-\delta)^+ = 0$, and hence $(u-\delta)^+ = 0$. Note that we have used the fact that the gradient of a Y_0 function f cannot vanish without f vanishing (cf. (4.67) below). The nonnegativity above follows from the inequalities,

$$\exp(u - v) - \exp(w - u) - k_1 \geq 2\sinh(u - \beta) - \sup_G k_1$$

$$\geq 2\sinh(\delta - \beta) - \sup_G k_1 \geq 2\sinh(\delta' - \beta) - \sup_G k_1 = 0.$$

Note that we have used the definitions of β, δ, and δ' above. In a similar way, one shows that $(u - \gamma)^- = \phi$ is the zero function, where $t^- = -(-t)^+$. The use of ϕ as a test function, coupled with the inequality,

$$\exp(u - v) - \exp(w - u) - k_1 \leq 0, \quad \text{on } \{u < \gamma\},$$

leads to this result. If we combine these two observations, we obtain the inequalities, (4.56, 4.57).

We now proceed to the equation itself. It is convenient to consider the problem of convex minimization, later shown to be equivalent. The reader may view this as a generalization of the classical Dirichlet principle. We thus consider,

$$\Phi(u) = \min_{L_2(G)} \Phi(f). \tag{4.60}$$

Here Φ is the convex functional defined by,

$$\Phi(f) =$$

$$\frac{1}{2}\int_G \epsilon|\nabla f|^2\, dx + \int_G F(f(x), x)\, dx - \int_G k_1(x) f(x)\, dx, \tag{4.61}$$

if $f \in \{f : F(f, \cdot) \in L_2(G)\} \bigcap \{f : (f - \bar{u}) \in Y_0\}$, and is defined to have the value ∞, for f not in this convex set. Here, F is an appropriate primitive:

$$F(s, x) =$$

$$\exp(-v(x)) \int_0^s \exp(\sigma)\, d\sigma + \exp(w(x)) \int_0^s -\exp(-\tau)\, d\tau. \tag{4.62}$$

Note that F is convex in s for each fixed x. In the terminology of [33], Φ is a proper convex functional on $L_2(G)$, hence has a finite minimum u under the following two sufficient conditions:

I Φ is coercive, i.e.,

$$\Phi(f)/\|f\|_{L_2} \to \infty \text{ as } \|f\|_{L_2} \to \infty;$$

II Φ is lower semicontinuous.

The reader will note that our characterization of coerciveness is more stringent than that given in [33]. This is due to the use of this condition when a convex functional is combined with a pseudomonotone operator in the sequel. We are able to verify the more stringent condition in the present context. In order to establish the lower semicontinuity of Φ, we make a preliminary observation about the functional $\int_G F$. Since F is continuous in its first argument, and is bounded from below by a constant,

$$F(s,x) \geq -\exp(-\alpha) - \exp(\beta), \tag{4.63}$$

it follows from Fatou's lemma of integration theory that

$$\liminf_{k\to\infty} \int_G F(f_k(x),x)\,dx \geq \int_G F(f(x),x)\,dx, \tag{4.64}$$

if $\{f_k\}$ is a sequence in L_2, pointwise convergent to f. The characterization of lower semicontinuity to be used is:

$$\forall a \in R^1,\ B = \{f \in L_2(G) : \Phi(f) \leq a\} \text{ is closed.} \tag{4.65}$$

Suppose, then, $f_k \to f$ in L_2, with $\Phi(f_k) \leq a$. Without loss of generality, by taking a subsequence if necessary, we may assume that f_k is pointwise a. e. convergent to f. By (4.61) and (4.63), we conclude that $\{f_k\}$ is bounded in H^1, where we have made use of (4.67). By the weak compactness of bounded closed sets in H^1, we conclude that $f_k \rightharpoonup f$ in H^1, where the indicated convergence is weak convergence. By the lower semicontinuity of the norm with respect to weak convergence, we conclude that

$$\liminf_{k\to\infty} \int_G \epsilon|\nabla f_k|^2 \geq \int_G \epsilon|\nabla f|^2. \tag{4.66}$$

Here, the norm employed is similar to that of (4.67) below. By (4.64, 4.66), and the convergence of $\{f_k\}$, we conclude that

$$\Phi(f) \leq$$

$$\liminf_{k\to\infty} \frac{1}{2}\int_G \epsilon|\nabla f_k|^2 + \liminf_{k\to\infty}\int_G F(f_k(x),x)\,dx - \int_G k_1(x)f(x)\,dx$$

$$\leq \liminf_{k\to\infty} \Phi(f_k) \leq a.$$

It follows that B is closed and Φ is lower semicontinuous. In order to demonstrate the coercivity of Φ, we make use of the norm, equivalent to that in H^1, given by

$$\|f\|^2 = \int_G |\nabla f|^2 + (\int_{\Sigma_D} \Gamma f)^2. \tag{4.67}$$

We recall in this connection that G is bounded, connected, and Σ_D has positive (relative) measure, which ensures the equivalence. A proof of the equivalence of (4.67) with the standard norm is demonstrated in ([67, p.

573]) by using standard facts documented in [97]. By estimating (4.61) in terms of (4.67), we infer, from the fixed boundary values of f on Σ_D in the case of finite $\Phi(f)$,

$$\liminf_{\|f\|\to\infty} \frac{\Phi(f)}{\|f\|^2} \geq \frac{1}{2} \inf \epsilon. \tag{4.68}$$

In bounding the limit infimum in (4.68), we have used (4.63), the domination of the L_2 norm by the H^1 norm, and the Schwarz inequality. If now, we form the quotient,

$$\frac{\Phi(f)}{\|f\|_{L_2}} = \frac{\Phi(f)}{\|f\|^2} \frac{\|f\|}{\|f\|_{L_2}} \|f\|,$$

we infer, from the domination of the L_2 norm by the H^1 norm, that Φ is coercive, and that a solution u of (4.60) exists. It remains to show that the minimization principle (4.60) implies (4.54). The technique is standard in the sense that one selects f of the form,

$$f = u + \eta\phi, \tag{4.69}$$

where η is an arbitrary real number, and ϕ is constrained to be in $Y_0 \bigcap L_\infty$. Usual techniques established in the calculus of variations yield (4.54) for ϕ so constrained. A density argument employing smooth functions gives the equation for all ϕ in Y_0. Uniqueness makes use of monotonicity of F in its first argument and the auxiliary norm, (4.67).

4.4 Admissible Lag and the Mapping $\mathbf{VW}_f$

In this section, we shall define the notion of admissible lagging, and use this to define in a careful way the map $\mathbf{VW}_f$. It would appear natural to consider the subsystem (4.72, 4.73) to follow, and attempt to define a joint system map from u to an image $[v, w]$, in terms of a weak solution of the system, and thus determine the map $\mathbf{VW}_f$. It does not appear, however, that such a correspondence need be unique in general, much less continuous. We are required to introduce, therefore, a family of such maps based on admissible lagging of terms in the current continuity subsystem, primarily in the recombination expression. More precisely, it is essential to consider dependence upon $[v, w] \in K$, as well as upon u. However, the lagging only marginally affects the issues of existence and maximum principles associated with the subsystem decoupling. Our approach is to separate out, in Definition 4.4.1, some minimal conditions needed for existence and maximum principles, but to defer the proofs, via variational inequalities, until the next two sections of this chapter. We then take up in Definition 4.4.2 the more comprehensive definition of admissible lagging, required for the additional properties of uniqueness and continuity, as a prelude to a well-defined subsystem map.

In this section, we assume that $\tilde{v}$, $\tilde{w}$, and $u = \mathbf{U}_f[\tilde{v}, \tilde{w}]$ are prescribed, as well as lagging in the recombination term of each equation of the current continuity subsystem, so that new functions, R_v and R_w, respectively, are obtained. This occurs through the substitution of $\tilde{v}$ and $\tilde{w}$ for v and w in selected instances of their occurrence.

Definition 4.4.1. *We assume for the form of R:*

$$R(u,v,w) = F(u,v,w)(\exp(w-v)-1), \tag{4.70}$$

where $F \geq 0$ *is assumed* C^∞ *on the closed set* $-\infty < u < \infty$, $\alpha \leq v \leq \beta$, $\alpha \leq w \leq \beta$. *The following critical assumptions are made concerning* R_v *and* R_w.

- *Within* R_v, *no lagging in* v *is permitted in the factor* $\exp(w-v)-1$, *and no lagging in* w *is permitted in the same factor in the function,* R_w.

In particular, the following relations on G are consequences of this assumption for $V, W \in K$:

$$\begin{aligned} R_v(u,V,W) &\geq 0, \text{ on } \{V=\alpha\}, \\ R_v(u,V,W) &\leq 0, \text{ on } \{V=\beta\}, \\ R_w(u,V,W) &\leq 0, \text{ on } \{W=\alpha\}, \\ R_w(u,V,W) &\geq 0, \text{ on } \{W=\beta\}. \end{aligned} \tag{4.71}$$

Thus, for purposes of eventually defining a map $\mathbf{VW}_f$, we are studying the subsystem,

$$-\nabla\cdot J_n - R_v = 0, \tag{4.72}$$
$$-\nabla\cdot J_p + R_w = 0. \tag{4.73}$$

Theorem 4.4.1. *The boundary-value problem associated with* (4.72, 4.73) *has a solution satisfying the maximum principles. Specifically, given* $[\tilde{v},\tilde{w}] \in K$, *and* $u = \mathbf{U}_f[\tilde{v},\tilde{w}]$, *with lagging satisfying Definition* 4.4.1, *then the coupled system* (4.72, 4.73), *subject to the Dirichlet boundary values on* Σ_D, *has a weak solution in* $K \cap X_u$ (cf. (4.82)). *(The precise result is stated in Lemma* 4.6.1.*)*

4.4.1 Admissible Lagging of the Continuity Subsystem

Definition 4.4.2. *Given* $[\tilde{v},\tilde{w}] \in K$ *and* $u = \mathbf{U}_f[\tilde{v},\tilde{w}]$, *an admissible lagging, yielding the system* (4.72, 4.73), *is required to satisfy the following conditions:*

- *Within* R_v, *no lagging in* v *is permitted in the factor* $\exp(w-v)-1$, *and no lagging in* w *is permitted in the same factor in the function,* R_w.
- *In the expressions for* J_n *and* J_p *in* (4.72) *and* (4.73), *the gradient multipliers* $\exp(-v)$ *and* $\exp(w)$ *may be replaced in tandem by the lagged variables* $\exp(-\tilde{v})$ *and* $\exp(\tilde{w})$.

– *The functions R_v and R_w of (u, v, w) are assumed to satisfy a pointwise semiboundedness condition in terms of the smallest eigenvalue of the energy functional. More precisely, for u as given above, and $[v_i, w_i] \in K$, $i = 1, 2$, we have*

$$\begin{aligned} &- [R_v(u, v_2, w_2) - R_v(u, v_1, w_1)](v_2 - v_1) \\ &+ [R_w(u, v_2, w_2) - R_w(u, v_1, w_1)](w_2 - w_1) \\ &\geq -\lambda_0(e^{\gamma-\beta}|v_2 - v_1|^2 + e^{\alpha-\delta}|w_2 - w_1|^2), \end{aligned} \tag{4.74}$$

where α and β are defined in (4.52, 4.53), *γ and δ are defined in* (4.56, 4.57), *and λ_0 is a nonnegative number less than the fundamental eigenvalue. Thus,*

$$\lambda_0 < \lambda_1, \tag{4.75}$$

where λ_1 is the positive infimum,

$$\lambda_1 = \inf_{z \neq 0} \int_G \rho |\nabla z|^2 / \int_G |z|^2. \tag{4.76}$$

In (4.76), *z is defined: $z \in H^1_\rho$, $\Gamma z \mid_{\Sigma_D} = 0$.*

Remark 4.4.1. Condition (4.74) is automatically satisfied when complete decoupling of the continuity equations occurs, provided R is also decreasing in v and increasing in w. Condition (4.74) is also satisfied in the important case when F in (4.70) is completely lagged, and its factor is completely unlagged. This yields a coupled system for which this framework is applicable, and monotonicity requirements are not present.

The existence of solutions of (4.103, 4.104), under the first assumption of Definition 4.4.2, has been noted in Theorem 4.4.1. Elementary modifications yield existence of the corresponding system when the gradient multipliers are lagged as well. No further assumptions are required.

In the next subsection, we shall address the issue of the definition of the family of maps $\mathbf{VW}_f$. Uniqueness of solutions of the system (4.103) and (4.104), or its counterpart in terms of lagged gradient multipliers, allows the association of such a solution with the image of the map $\mathbf{VW}_f$. It will not be necessary to make any further assumptions beyond those of Definition 4.4.2.

– Uniqueness of images of $\mathbf{VW}_f$, and hence the well-definedness property of this map, as well as the continuous dependence of images of this map upon the L_2 gradient norm of u and the L_2 norms of the lagged variables, follow from the definition of admissible lagging in Definition 4.4.2. The arguments require case distinctions based on whether $\exp(-v)$ and $\exp(w)$ are lagged as multipliers of the gradient terms of the current continuity subsystem. These issues are addressed in Lemmas 4.4.2 and 4.7.2 below.

4.4.2 Uniqueness and Definition of the Map $\mathbf{VW}_f$

We begin with some critical identities. The proofs are by inspection.

Lemma 4.4.1. *Let $[\tilde{v}_1, \tilde{w}_1]$ and $[\tilde{v}_2, \tilde{w}_2]$ be lagged variables in K, and let u_1 and u_2 be the images under the map $\mathbf{U}_f$ of these pairs. Let $[v_1, w_1]$ and $[v_2, w_2]$ be the corresponding solution pairs of the current continuity subsystems. These are defined by (4.103, 4.104) in the case where the gradient multiplier terms are not lagged, and by the correspondingly modified system otherwise. Then the following identities, expressed in terms of duality pairings, hold:*

– *If lagging occurs in the gradient terms, then*

$$\begin{aligned}\langle \mu_n(\nabla u_1)\exp(u_1 - \tilde{v}_1)\nabla(v_1 - v_2), \nabla(v_1 - v_2)\rangle &= \\ -\langle \mu_n(\nabla u_1)[\exp(u_1 - \tilde{v}_1) - \exp(u_2 - \tilde{v}_2)]\nabla v_2, \nabla(v_1 - v_2)\rangle \\ -\langle [\mu_n(\nabla u_1) - \mu_n(\nabla u_2)]\exp(u_2 - \tilde{v}_2)\nabla v_2, \nabla(v_1 - v_2)\rangle \\ +\langle R_v(u_1, v_1, w_1) - R_v(u_2, v_2, w_2), v_1 - v_2\rangle, \end{aligned} \tag{4.77}$$

$$\begin{aligned}\langle \mu_p(\nabla u_1)\exp(\tilde{w}_1 - u_1)\nabla(w_1 - w_2), \nabla(w_1 - w_2)\rangle &= \\ -\langle \mu_p(\nabla u_1)[\exp(\tilde{w}_1 - u_1) - \exp(\tilde{w}_2 - u_2)]\nabla w_2, \nabla(w_1 - w_2)\rangle \\ -\langle [\mu_p(\nabla u_1) - \mu_p(\nabla u_2)]\exp(\tilde{w}_2 - u_2)\nabla w_2, \nabla(w_1 - w_2)\rangle \\ -\langle R_w(u_1, v_1, w_1) - R_w(u_2, v_2, w_2), w_1 - w_2\rangle. \end{aligned} \tag{4.78}$$

– *If lagging does not occur in the gradient terms, then, with $*$ designating averages, we have*

$$\begin{aligned}\langle \mu_n^* \exp(u^* - v^*)\nabla(v_1 - v_2), \nabla(v_1 - v_2)\rangle &= \\ \langle \mu_n^* \exp(u^* - v^*)\nabla(v_1 - v_2), \nabla(u_1 - u_2)\rangle \\ -\langle [\mu_n(\nabla u_1) - \mu_n(\nabla u_2)]\exp(u^* - v^*)\nabla v^*, \nabla((v_1 - v_2) - (u_1 - u_2))\rangle \\ +\langle R_v(u_1, v_1, w_1) - R_v(u_2, v_2, w_2), v_1 - v_2\rangle, \end{aligned} \tag{4.79}$$

$$\begin{aligned}\langle \mu_p^* \exp(w^* - u^*)\nabla(w_1 - w_2), \nabla(w_1 - w_2)\rangle &= \\ \langle \mu_p^* \exp(w^* - u^*)\nabla(w_1 - w_2), \nabla(u_1 - u_2)\rangle \\ -\langle [\mu_p(\nabla u_1) - \mu_p(\nabla u_2)]\exp(w^* - u^*)\nabla w^*, \nabla((w_1 - w_2) - (u_1 - u_2))\rangle \\ -\langle R_w(u_1, v_1, w_1) - R_w(u_2, v_2, w_2), w_1 - w_2\rangle. \end{aligned} \tag{4.80}$$

Lemma 4.4.2. *The map $\mathbf{VW}_f$ is well defined for any admissible lagging as described in Definition 4.4.2. In particular, for $[\tilde{v}, \tilde{w}] \in K$ and $u = \mathbf{U}_f[\tilde{v}, \tilde{w}]$, there is a unique pair $[V, W] = \mathbf{VW}_f[u, \tilde{v}, \tilde{w}]$ satisfying the system (4.103, 4.104), subject to the stated boundary conditions, as specified by (4.84).*

Proof. Take $u_1 = u_2 = u$, $\tilde{v}_1 = \tilde{v}_2 = \tilde{v}$, and $\tilde{w}_1 = \tilde{w}_2 = \tilde{w}$. In the case where lagging occurs in the gradient terms, addition of (4.77) and (4.78), and an application of (4.74), yield

$$\begin{aligned} \langle \mu_n \exp(u - \tilde{v})\nabla(v_1 - v_2), \nabla(v_1 - v_2)\rangle &\quad + \\ \langle \mu_p \exp(\tilde{w} - u)\nabla(w_1 - w_2), \nabla(w_1 - w_2)\rangle &\quad \leq \\ \lambda_0(e^{\gamma-\beta}\|v_1 - v_2\|^2_{L_2} + e^{\alpha-\delta}\|w_1 - w_2\|^2_{L_2}), & \end{aligned} \tag{4.81}$$

where λ_0 satisfies (4.75) in terms of the fundamental eigenvalue λ_1. A direct estimation, making use of (4.76), the definitions of α, β, γ, δ, and λ_1, and Assumption 4.4, yields an inequality of opposite sense to (4.81), with $\lambda_0 \to \lambda_1$. This can only occur if the norms of $v_1 - v_2$ and $w_1 - w_2$ in H^1_ρ are 0; this proves uniqueness in the case of gradient term lagging. The case of no such gradient term lagging is even simpler, via (4.79) and (4.80). In this case, we need make use only of the relation, $u_1 = u_2 = u$.

4.5 A Variational Inequality for the Current Continuity Subsystem

In section §4.4, we introduced the subsystem,

$$\begin{aligned} -\nabla\cdot \mathbf{J}_n - R_v &= 0, \\ -\nabla\cdot \mathbf{J}_p + R_w &= 0, \end{aligned}$$

as equations (4.72) and (4.73). In order to establish the existence of weak solutions of this subsystem, where boundary conditions are imposed as previously, we shall use different arguments than those based upon the minimization of convex functionals. This is necessitated because the subsystem is possibly *coupled*, and is not of gradient type in general. Furthermore, the existence alone of a solution of the subsystem is not acceptable without the derivation of maximum principles. In this transition section, then, we shall examine a variational inequality version of (4.72) and (4.73). The structure of the argument will follow in a general way the lines of [67, Sec. 5]. The variational inequality will be defined on a mobility weighted Sobolev space product, or rather, an appropriate convex subset thereof, as dictated by boundary conditions and maximum principles, which are imposed as "obstacles". The reader may recall the simplest occurrence of variational inequalities, as resulting from the determination of an element y of minimal norm in a closed convex subset of a Hilbert space. In that case, the inequality expresses a geometric condition, describing the angle between y and $z - y$, for any z in the convex set. Here the situation is more complicated, since minimization does not play a role. Now, it will be essential to embed the mobility weighted spaces in the standard weighted Sobolev spaces, introduced earlier in §4.2. The solution of the variational inequality will thus lie in the Hilbert product space,

$$X_u = H^1_{\mu_n}(G) \times H^1_{\mu_p}(G). \tag{4.82}$$

Note that we have used the subscripts, μ_n and μ_p, to denote the spaces of locally square integrable functions for which (4.49) is finite, with ρ replaced by the respective mobilities. Within this product space, the variational inequality is taken over the closed convex set,

$$K_0 = K \bigcap K_{aff}, \tag{4.83}$$

where

$$K_{aff} = \{[v, w] \in X_u : \Gamma([v, w] - [\bar{v}, \bar{w}]) \mid_{\Sigma_D} = 0\}. \tag{4.84}$$

Here, K has been defined above in (4.51). The reader will notice that both the maximum principles and the boundary conditions have been incorporated into the definition of K_0. The variational inequality to be verified in this section assumes the form given by (4.95) to follow, where the individual terms are described by (4.87) and (4.88).

4.5.1 Abstract Inequality Formulation

The use of variational inequalities is based upon the following result (cf. [18] and [64, Proposition 3.1.5]), *stated* here in a duality pairing formulation.

Lemma 4.5.1. *Let $\mathcal{K}_0$ be a closed, convex subset of a reflexive Banach space $\mathcal{X}$, of the form $\mathcal{K}_0 = z_0 + \mathcal{K}$ with $0 \in \mathcal{K}$, suppose that Φ is a proper, convex, lower semicontinuous functional defined on $\mathcal{K}_0$, $\Phi(z_0) = 0$, and that $B : \mathcal{K}_0 \to \mathcal{X}^*$ is pseudomonotone, i.e.,*

1. *$\forall z$ in $\mathcal{X}$, $\langle B(y), y - z\rangle \leq \liminf_{i\to\infty} \langle B(y_i), y_i - z\rangle$* whenever

$$y_i \rightharpoonup y \text{ (weakly) and } \limsup_{i\to\infty} \langle B(y_i), y_i - y\rangle \leq 0,$$

for $y_i, y \in \mathcal{K}_0$; and,

2. *$B(S)$ is bounded in $\mathcal{X}^*$, if S is bounded in $\mathcal{X}$.*

Suppose that B and Φ jointly define a coercive functional, i.e., there exist a positive number, λ, and a monotone, nondecreasing function, $\omega : [\lambda, \infty) \to [0, \infty)$, $\omega^{-1}(t)$ a bounded set for each t, such that

$$\frac{\langle B(z), z - z_0\rangle + \Phi(z)}{\|z - z_0\|} \geq \omega(\|z - z_0\|), \quad \|z - z_0\| \geq \lambda.$$

Then, for each $b \in \mathcal{X}^*$, the variational inequality,

$$\langle B(y) - b, z - y\rangle + \Phi(z) - \Phi(y) \geq 0, \ \forall z \in \mathcal{K}_0, \tag{4.85}$$

possesses a solution $y \in \mathcal{K}_0$. Furthermore, the following bound holds for such a solution:

$$\|b\| \leq \sigma \in \mathit{Range}(\omega) \Rightarrow \|y - z_0\| \leq \sup \omega^{-1}(\sigma). \tag{4.86}$$

The hypothesis,

$$0 \in \mathcal{K},$$

in the statement of Lemma 4.5.1 is due to the slightly more general form given here. Indeed, the form referenced in [64] is the special case, $z_0 = 0$. Note also that the hypothesis made there,

$$\Phi(0) = 0,$$

is replaced here by

$$\Phi(z_0) = 0.$$

Its use allows the precise estimate (4.86). The reduction to the special case of [64] is easily achieved by use of the translation map,

$$\tau(z) = z - z_0,$$

whereby the action of mappings on $\mathcal{K}_0$ is reduced to action on $\mathcal{K}$.

Remark 4.5.1. There are three steps in the program to establish the existence of the solution of the coupled subsystem (4.72, 4.73). Two of these steps are carried out in this section, beginning with this subsection, and the remaining one in the following section, §4.6. They are:

1. The identification of the elements in Lemma 4.5.1;
2. The verification of the hypotheses of this lemma, and, hence the existence of a solution of the corresponding variational inequality;
3. The verification that the variational inequality is equivalent to the weak form of subsystem (4.72, 4.73). The test functions in such a weak formulation comprise pairs $[\phi, \psi]$ in X_u, such that the components have zero trace on Σ_D.

For the appropriate identifications, set $b = 0$, $z_0 = [\bar{v}, \bar{w}]$, $\mathcal{X} = X_u$, and $\mathcal{K}_0 = K_0$ (cf. (4.82, 4.83)). To continue the identifications, we define $\Phi = \Psi_u$, $B = A_u$, where Ψ_u and A_u are defined on K_0, by the following relations. Given $u \in \text{Range}(\mathbf{U}_f)$, set

$$\Psi_u(v, w) = \int_G \mu_n e^{u-v} |\nabla v|^2 + \int_G \mu_p e^{w-u} |\nabla w|^2 + C, \tag{4.87}$$

where C is chosen so that $\Psi_u(\bar{v}, \bar{w}) = 0$. The latter is a technical hypothesis discussed above. Similarly, for $u \in \text{Range}(\mathbf{U}_f)$, $A_u : K_0 \to X_u^*$ is defined by

$$\begin{aligned}
\langle A_u(V, W), (v, w) \rangle &= \\
\langle A_{1,u}(V, W), v \rangle &+ \langle A_{2,u}(V, W), w \rangle, && (4.88) \\
\langle A_{1,u}(V, W), v \rangle &= -2 \int_G R_v(u, V, W) v, && (4.89) \\
\langle A_{2,u}(V, W), w \rangle &= 2 \int_G R_w(u, V, W) w. && (4.90)
\end{aligned}$$

Here, $[V, W] \in K_0$ and $[v, w] \in X_u$; also, the terms on the right hand sides of (4.89, 4.90) are properly signed recombination/generation terms as discussed earlier in this section in Definition 4.4.1.

4.5.2 The Concrete Variational Inequality

In this subsection, we shall verify the hypotheses of the abstract Lemma 4.5.1, and draw the desired conclusion in terms of the inequality derived from the current continuity relations, subject to the obstacles induced by the maximum principles. The following lemma incorporates these steps.

Lemma 4.5.2. *X_u is a Hilbert space, continuously embedded in $\prod_1^2 H_\rho^1(G)$, and hence continuously embedded in $\prod_1^2 W_q^1(G)$ for*

$$1 \leq q < \frac{2N}{N+\theta}. \tag{4.91}$$

Note that θ has been introduced in Assumption 4.4. It follows that X_u is compactly embedded in $\prod_1^2 L_r(G)$ for

$$1 \leq r < \frac{2N}{N+(\theta-2)}, \tag{4.92}$$

and also that the set K_0 is closed and convex in X_u. A norm equivalent to that induced by (4.82) is taken to be

$$\|[v,w]\|^2_{X_u} \quad = \tag{4.93}$$

$$\int_G (\mu_n e^{u-\beta}|\nabla v|^2 + \mu_p e^{\alpha-u}|\nabla w|^2) \quad + \quad (\int_{\Sigma_D} \Gamma v)^2 + (\int_{\Sigma_D} \Gamma w)^2.$$

Furthermore, the functional Ψ_u is proper, convex, and lower semicontinuous on K_0, and vanishes at $[\bar{v},\bar{w}]$, while the map A_u is pseudomonotone. Moreover, A_u and Ψ_u jointly define a coercive functional, in the sense of the preceding lemma, with respect to the function,

$$\omega(t) = \frac{1}{2}(t-\lambda), \quad t \geq \lambda. \tag{4.94}$$

Here, λ is specified below in (4.102). In particular, the variational inequality,

$$\langle A_u(V,W), [v,w]-[V,W]\rangle \ +$$

$$\Psi_u(v,w) - \Psi_u(V,W) \geq 0, \quad \forall [v,w] \in K_0, \tag{4.95}$$

has a solution $[V,W] \in K_0$, and the ' priori' bound,

$$\|[V,W]\|_{X_u} \leq \lambda, \tag{4.96}$$

holds. Note that λ is independent of u.

Proof. The continuous embedding property, satisfied by X_u, follows directly from (4.44) and (4.45) in the original topology of X_u (cf. (4.82)). The norm equivalence of (4.93) is clear; it will be demonstrated that X_u is a Hilbert space in this norm. Thus, let f_k be a Cauchy sequence in (4.93); this sequence has a limit f in the Hilbert space $\prod_1^2 H_\rho^1$, via the domination of the product norm on $\prod_1^2 H_\rho^1$ by (4.93). We need only show that $\|f\|_{X_u}$ is finite. By the triangle inequality, we observe that $\lim_{k\to\infty} \|f_k\|_{X_u}$ exists, and is equal to a nonnegative real number, say, ζ. It suffices to show that

$$\int_{\mathcal{U}} (\mu_n \exp(u-\beta)|\nabla v|^2 + \mu_p \exp(\alpha - u)|\nabla w|^2) \\ + (\int_{\Sigma_D} \Gamma v)^2 + (\int_{\Sigma_D} \Gamma w)^2 \le \zeta^2, \tag{4.97}$$

for each open subset $\mathcal{U}$ of G with compact closure in G. Here, we have used the notation, $[v, w] = f$. By Assumption 4.4 above (cf. (4.44, 4.45)), we see that (4.97) is a consequence of the $H^1(\mathcal{U})$ seminorm convergence of f_k. It follows that X_u is a Hilbert space. The embedding property described in the final section by (4.114) yields (4.91) for q as stated. By use of this value of q, the compactness result, for r as described by (4.92), follows from [1, p. 144]. We can now conclude that the convex set $K \bigcap X_u$ is closed; one uses the pointwise subsequential convergence of an L_q convergent sequence, $q < \infty$, to deduce this. The embeddings, and the continuity of the trace map (cf. inequality (4.115) of the final section), allow one to complete the argument that K_0 is closed. Ψ_u is convex, as the sum of convex functionals; each of the nonconstant terms is the composition of a convex functional with a convex exponential function, hence is convex as stated. The proof that Ψ_u is lower semicontinuous on K_0 uses the argument of Lemma 4.3.1 and uses the fact that, on K_0, the (squared) leading seminorm part of (4.93) is equivalent to $\Psi_u - C$. The remaining nontrivial properties to verify are the pseudomonotonicity of A_u, and the coercivity property. For the former, the boundedness of A_u as a mapping into X_u^* results from the uniform pointwise boundedness of the functions $R_v(u, V, W)$ and $R_w(u, V, W)$ for arbitrary $[V, W]$ in K_0. This pointwise bound permits a simple estimate for the functional of (4.88) in terms of an L_1 estimate, and hence an X_u estimate, on $[v, w]$. For the other property of pseudomonotonicity, suppose that $f_k \rightharpoonup f$ (weakly) in X_u, where $f_k = [v_k, w_k]$ and $f = [v, w]$. Without requiring the lim sup property in the characterization of pseudomonotonicity, we shall show that, for each $g \in X_u$,

$$\langle A_u(f), f - g\rangle = \lim_{k\to\infty} \langle A_u(f_k), f_k - g\rangle, \tag{4.98}$$

which is a strengthened version of the property. Now select $q > 1$ as in (4.114); as follows from (4.92), with $r = q$, the embedding, $X_u \hookrightarrow \prod_1^2 L_q(G)$, is compact. In particular, the sequence $\{f_k\}$ is convergent to f in this product Lebesgue space, since a compact mapping takes weakly convergent sequences into convergent sequences. The verification of (4.98) is established, via Hölder's inequality, by use of the L_q convergence of f_k and the following convergence result:

– For q' satisfying $\frac{1}{q} + \frac{1}{q'} = 1$, we have convergence in $L_{q'}$ given by

$$\lim_{k \to \infty} R_v(u, v_k, w_k) = R_v(u, v, w) \tag{4.99}$$

with a similar relation for R_w.

The property (4.99) is a consequence of the theory of superposition operators as presented in [91, Chap. 18]. The uniform pointwise boundedness of the functions R_v, together with the local Lipschitz property in the variables v and w, implies (in particular) uniform continuity of the mapping,

$$R_v(u, \cdot, \cdot) : [v, w] \in \prod_1^2 L_p \mapsto R_v(u(s), v(s), w(s)) \in \prod_1^2 L_{p'}.$$

The estimate is straightforward, and can be carried out independently of reference to [91], where only one variable v is considered. This now yields (4.99), and directly the pseudomonotonicity.

We are now left with proving the coercivity. By the definition of the X_u norm and that of Ψ_u on K_0, and by the boundedness of A_u, we obtain the relation,

$$\frac{\langle A_u(f), f - f_0 \rangle + \Psi_u(f)}{\|f - f_0\|_{X_u}} \geq \frac{1}{2}\|f - f_0\|_{X_u} - c, \tag{4.100}$$

provided, for example,

$$\frac{\|f_0\|_{X_u}}{\|f\|_{X_u}} \leq \frac{1}{2}, \tag{4.101}$$

where c is a positive constant not depending on u, and f is otherwise unrestricted within K_0. Here, we have used the symbol f_0 for $[\bar{v}, \bar{w}]$. Also, it is not necessary that $f_0 \neq 0$, and (4.101) is vacuous in this case. By defining

$$\lambda = \max(2c, 2\|f_0\|_{X_u}), \tag{4.102}$$

we obtain the desired function ω described in the statement of the lemma.

4.6 Equivalence with the Current Continuity Subsystem

The final step in establishing the existence of a solution of the current continuity subsystem, (4.72, 4.73), is to show that a solution of the variational inequality (4.95) satisfies this subsystem in a weak sense.

We are now prepared to state the result which forms the basis for the definition of the map $\mathbf{VW}_f$.

Lemma 4.6.1. *Any solution $[V, W]$ of the variational inequality* (4.95) *satisfies the (lagged) subsystem* (4.72, 4.73) *in a weak sense:*

$$\langle \mu_n \exp(u - V)\nabla V, \nabla\psi\rangle - \langle R_v(u, V, W), \psi\rangle = 0, \tag{4.103}$$

$$\langle \mu_p \exp(W - u)\nabla W, \nabla\omega\rangle + \langle R_w(u, V, W), \omega\rangle = 0, \tag{4.104}$$

for all $[\psi, \omega]$ in $Y_{u,0}$; the latter test function pair in X_u has zero trace on Σ_D.

Proof. We shall show that (4.103) holds; an analogous argument yields (4.104). The proof is simplified considerably if we consider the more restricted case when the exponential multipliers in the gradient parts of the system equations are lagged. To make the proof more accessible, we present the details in this case. The arguments presented are readily adapted to the system (4.103) and (4.104) by use of the Slotboom variables; we shall indicate the modifications at the appropriate time. Thus, suppose that $\exp(-V)$ is replaced by $\exp(-v_*)$ in (4.103) as an appropriate "lag". By taking limits, we may assume that ψ is pointwise bounded; by normalization we may assume that ψ is pointwise bounded by 1 and has norm in $H^1_{\mu_n}$ not exceeding $e^{-|\alpha-\delta|}$. Given $\{\eta_k\}$, $\eta_k > 0$, $\eta_k \to 0$, we select sequences $\{\alpha_k\}$ and $\{\beta_k\}$ such that each of the following three conditions is satisfied:

1. $\alpha_k - \alpha := \epsilon_k = \beta - \beta_k < \eta_k e^{|\delta-\alpha|}$.
2. For $\mathcal{A}_k = \{\alpha < V < \alpha_k\} \subset \mathbb{R}^N$, $\mathcal{B}_k = \{\beta_k < V < \beta\} \subset \mathbb{R}^N$, require
$$\int_{\mathcal{A}_k \cup \mathcal{B}_k} \mu_n |\nabla V|^2 < \left(\frac{\eta_k}{4}\right)^2.$$
3. measure $(\mathcal{A}_k \cup \mathcal{B}_k) \sup |P| < \frac{\eta_k}{4}$.

Here we have employed the substitution, $P = R_v$, and have used a theorem of Lebesgue on convergent sums of integrals over disjoint sets to obtain item two. In order to proceed from the variational inequality (4.95) to (4.103), we define the "cutoff" of $V \pm \epsilon_k\psi$:

$$v = v_\pm = (V \pm \epsilon_k\psi - \alpha)^+ + \alpha + (\beta - V \mp \epsilon_k\psi)^-.$$

Note that $[v_+, w]$ and $[v_-, w]$ are legitimate members of K_0 when the choice $w = W$ is made. We observe that

$$v_\pm - V = \pm\epsilon_k\psi \text{ on } \{\alpha_k \le V \le \beta_k\}, \tag{4.105}$$

$$v_+ - V = \epsilon_k\psi \text{ on } \{V = \alpha\} \cap \{\psi \ge 0\}, \tag{4.106}$$

$$v_- - V = -\epsilon_k\psi \text{ on } \{V = \alpha\} \cap \{\psi \le 0\}, \tag{4.107}$$

$$v_+ - V = \epsilon_k\psi \text{ on } \{V = \beta\} \cap \{\psi \le 0\}, \tag{4.108}$$

$$v_- - V = -\epsilon_k\psi \text{ on } \{V = \beta\} \cap \{\psi \ge 0\}. \tag{4.109}$$

We require some notation for sets. Thus, make the substitutions,

$$\mathcal{D}_k = \{\alpha_k \le V \le \beta_k\},$$

$$\mathcal{E}_k = \mathcal{D}_k \cup \{V = \alpha\} \cup \{V = \beta\},$$

$$\mathcal{F}_k = \mathcal{D}_k \cup (\{V = \alpha\} \cap \{\psi \ge 0\}) \cup (\{V = \beta\} \cap \{\psi \le 0\}).$$

Substitution of $[v_+, w = W]$ into (4.95) yields, after division by $2\epsilon_k$,

$$\int_{\mathcal{E}_k} \mu_n \exp(u - v_*) \nabla V \cdot \nabla \psi - \int_{\mathcal{F}_k} P\psi \ge$$

$$- \int_{\mathcal{A}_k} |\mu_n \exp(u - v_*) \nabla V \cdot \nabla \psi| - \int_{\mathcal{B}_k} |\mu_n \exp(u - v_*) \nabla V \cdot \nabla \psi|$$

$$- \epsilon_k/2 \int_G \mu_n \exp(u - v_*) |\nabla \psi|^2$$

$$+ \epsilon_k^{-1} \int_{\mathcal{A}_k} P(v_+ - V) + \epsilon_k^{-1} \int_{\mathcal{B}_k} P(v_+ - V). \tag{4.110}$$

In computing the set over which the second integral on the left hand side of this inequality is taken, we have used the fact that $v_+ - V$ vanishes on the two sets, $\{V = \alpha\} \cap \{\psi \le 0\}$ and $\{V = \beta\} \cap \{\psi \ge 0\}$. At this point in the proof, we employ the critical properties that $P \ge 0$ on $\{V = \alpha\}$ and $P \le 0$ on $\{V = \beta\}$, as described in (4.71); these inequalities are at the heart of the equivalence. This permits us to add two key terms involving integrals of $-P\psi$, without changing the sense of the above inequality. These are terms involving integration over $\{V = \alpha\} \cap \{\psi \le 0\}$ and $\{V = \beta\} \cap \{\psi \ge 0\}$. When these integrals are added, and the domain of integration for the second left hand side integral is consolidated, we may rewrite the left hand side of the inequality as

$$\int_{\mathcal{E}_k} \mu_n \exp(u - v_*) \nabla V \cdot \nabla \psi - \int_{\mathcal{E}_k} P\psi.$$

The right hand side of (4.110) is at least $-\eta_k$, by the restrictions imposed on α_k, β_k, and ϵ_k. More precisely, we have obtained the inequality,

$$\int_{\mathcal{E}_k} \mu_n \exp(u - v_*) \nabla V \cdot \nabla \psi - \int_{\mathcal{E}_k} P\psi > -\eta_k. \tag{4.111}$$

The companion inequality, whereby the left hand side is shown to be bounded above by η_k, is obtained by substitution of v_-. In fact, the inequality,

$$\int_{\mathcal{E}_k} \mu_n \exp(u - v_*) \nabla V \cdot \nabla \psi - \int_{\mathcal{E}_k} P\psi < \eta_k, \tag{4.112}$$

is obtained as a result of this substitution, and use of arguments parallel to those above. It follows that the common left hand side of (4.111) and (4.112) has zero limit as $k \to \infty$. The three inequalities satisfied by $\{\eta_k\}$ above then demonstrate that this zero limit is

$$\int_G \mu_n \exp(u - v_*)\nabla V \cdot \nabla\psi - \int_G P\psi.$$

The case where the exponential gradient multipliers are not lagged is handled by setting $v_* = 0$ and using the variable $\exp(-V)$ as the basic variable, while constructing parallel arguments. The pointwise intervals must be accordingly shifted to account for the Slotboom transformation.

4.7 Compactness and Continuity of $\mathbf{VW}_f$ and Fixed Points of $\mathbf{T}_f$

In order to guarantee that we can work within the framework of L_2 product spaces in the final subsection of this section, when we consider the composition of $\mathbf{U}_f$ and $\mathbf{VW}_f$, we shall make a more stringent assumption upon r, introduced in (4.92); specifically, we require that $r \geq 2$. This leads to the following assumption, maintained throughout this section.

Assumption 4.6. It is assumed that

$$\theta < 2. \tag{4.113}$$

In particular, $\frac{2N}{N+(\theta-2)} > 2$.

Here, θ was originally introduced in Assumptions 4.4 and 4.5, and the value assumed in (4.113) is consistent with known rates discussed earlier.

4.7.1 Compactness

Compactness estimates are cited in Lemma 4.7.1 to follow. This is a brief subsection, highlighting this essential property.

Lemma 4.7.1. *Given $[\tilde{v}, \tilde{w}]$ in K and its image u under $\mathbf{U}_f$, there is an 'a priori' bound λ on the $\prod_1^2 H_\rho^1(G)$ norm of the image point of the map $\mathbf{VW}_f$, which is independent of $[\tilde{v}, \tilde{w}]$. In particular, the range of the map $\mathbf{VW}_f$ is relatively compact in $\prod_1^2 L_2(G)$.*

Proof. The result, in the case where no lagging occurs in the gradient terms, is a consequence of (4.96), as well as the embedding properties described in Lemma 4.5.2. The other case is virtually identical.

4.7.2 Continuity

The continuity required of $\mathbf{VW}_f$, for the existence of fixed points of the Gummel map as presented in the next subsection, is product L_2 continuity in the dependence upon $\tilde{v}$ and $\tilde{w}$. In fact, a significantly stronger continuity result is valid when the range is normed by the weighted energy space norm.

Since the weaker result is straightforward, we shall present it and its proof here, and shall include the stronger result, together with its proof, in the final section as Lemma 4.8.2. That proof is more computational, and depends on direct estimation. Some readers may prefer it to the proof of Lemma 4.7.2 below, however.

Lemma 4.7.2. *The mapping* $\mathbf{VW}_f$ *is continuous, in its* $L_2(G)$ *dependence upon* $\tilde{v}$ *and* $\tilde{w}$, *and in its* $H^1_\rho(G)$ *dependence upon* u, *as a mapping into* $\prod_1^2 L_2(G)$.

Proof. Because of the compact embedding described in the previous subsection, strong sequential continuity of $\mathbf{VW}_f$, as a mapping into $\prod_1^2 L_2$, is implied by weak sequential convergence in $\prod_1^2 H^1_\rho$. This is equivalent to showing that the corresponding weak solution systems, as formulated in (4.103) and (4.104), converge, as systems, to the particular system taken at the function pair at which continuity is verified. Moreover, uniqueness allows this weak convergence to be demonstrated subsequentially. The equivalence of weak convergence with the system convergence, as just stated, is a simple consequence of the relation (4.99), for the case $q = q' = 2$; the relation (4.99) is used a second time in demonstrating the convergence (or subsequential convergence) of the systems themselves. This leaves only the convergence of the leading (gradient) parts of (4.103, 4.104) to be demonstrated. Without loss of generality, we may assume smooth test functions. Given a sequence $\{[\tilde{v}_j, \tilde{w}_j]\}$ in K, convergent in the product L_2 metric, let u denote the image of its limit under $\mathbf{U}_f$. By the usual criterion, we need only show that every subsequence of the image sequence $\{[v_j, w_j]\}$, has a further subsequence convergent to an invariant limit. Given such a subsequence, select a further subsequence, conveniently denoted $\{[v_{j_k}, w_{j_k}]\}$, which is weakly convergent in $\prod_1^2 H^1_\rho$. Using the well known fact that the corresponding weak convergence is not altered if the integrands are multiplied by a boundedly convergent sequence, and taking a further subsequence if necessary, we conclude that the integral terms,

$$\langle \mu_n(\nabla u) \exp(u_{j_k} - v_{j_k}) \nabla v_{j_k}, \nabla \psi \rangle,$$

and

$$\langle \mu_p(\nabla u) \exp(w_{j_k} - u_{j_k}) \nabla w_{j_k}, \nabla \omega \rangle,$$

are convergent to the expected limit. A routine application of the triangle inequality, making use of the terms just described, as well as terms estimated by the bounded pointwise convergence of $\mu_n(\nabla u_{j_k})$ to $\mu_n(\nabla u)$ (and similarly for μ_p), yields the convergence of the leading parts as required. In this connection, the local pointwise convergence of ∇u_{j_k} is converted to global pointwise convergence by a diagonalization procedure. This concludes the proof of continuity.

4.7.3 The Gummel Map and its Fixed Points

In §4.3 we defined the map $\mathbf{U}_f$, and in §4.4 we defined the map $\mathbf{VW}_f$.

Definition 4.7.1 (Gummel map). *By the Gummel map, $\mathbf{T}_f$, we mean any composition, $\mathbf{VW}_f \circ \mathbf{U}_f$, of the maps just cited. In particular, $\mathbf{T}_f$ is not a single map, but one of a family of maps.*

According to the theory of this chapter, $\mathbf{T}_f$ is a mapping of the closed, convex set K into itself. Moreover, under the hypothesis on θ, $\mathbf{T}_f$ is continuous as a mapping of $\prod_1^2 L_2(G)$ into itself, and has relatively compact range. It follows that $\mathbf{T}_f$ has at least one fixed point in K as a consequence of the Schauder fixed point theorem.

We have the following existence result.

Theorem 4.7.1. *Under Assumptions* 4.5 and 4.6, *there exists a weak solution of the drift-diffusion system. Such a solution may be identified with a fixed point of any one of the family of maps $\mathbf{T}_f$ defined above. This family is not vacuous (cf. the discussion in Remark* 4.4.1, *following Definition* 4.4.2*).*

4.8 Technical Properties of Norms and Mappings

We treat the two issues of norm equivalence and continuity in this section. We begin with the first.

4.8.1 Norm Equivalence

Our first result deals with norm equivalence as asserted in §4.2.

Lemma 4.8.1. *The norms* (4.49) *and* (4.50) *are equivalent norms.*

Proof. The subcase that $\theta = 0$ is the trivial subcase in what follows. In this instance,

$$H_\rho^1(G) \equiv W_2^1(G).$$

Now, it is shown in [92] that the embedding,

$$H_\rho^1(G) \hookrightarrow W_q^1(G), \tag{4.114}$$

is continuous for $0 < \theta < N$ and $1 \le q < \frac{2N}{N+\theta}$. To show that (4.49) defines a norm equivalent to that in $H_\rho^1(G)$, as described in (4.50), recall the trace inequality (cf. [1]),

$$\|f\|_{L_r(\Sigma_D)} \le C_0 \|f\|_{W_q^1(G)}, \tag{4.115}$$

for $q < N$ and

$$q \le r \le \frac{(N-1)q}{N-q}.$$

The conjunction of (4.114) and (4.115), with $q = r = 1$, yields the domination of (4.49) by the standard norm. By the open mapping theorem of functional analysis, the norm equivalence will follow if the completeness of $H^1_\rho(G)$ in the norm (4.49) can be demonstrated. Thus, given a Cauchy sequence f_k in this norm, there is a function f such that

$$\int_G \rho|\nabla(f - f_k)|^2 \to 0. \tag{4.116}$$

This follows from a nontrivial equivalence of the coset norm on

$$H^1_\rho(G)/\{\text{constants}\},$$

with the seminorm induced by (4.116). The proof of this fact uses ideas contained in [64, Sec. 4.1], particularly a version of Corollary 4.1.4, adjusted for weighted spaces. The adjustment requires a routine interpretation of Young's inequality for weighted spaces. Returning to the completeness argument, we shall select a constant b such that

$$f_k \to f + b,$$

in the norm of (4.49). Define $b_1 = \lim_{k\to\infty} \int_{\Sigma_D} \Gamma f_k$, $b_2 = \int_{\Sigma_D} \Gamma f$. Then select b to satisfy

$$b_1 = b_2 + b \,\text{measure}(\Sigma_D).$$

With these definitions, and the convergence result (4.116), we deduce that $f_k \to f + b$ in the norm (4.49). This proves completeness, and hence the desired equivalence of norms.

4.8.2 Enhanced Continuity for the Subsystem Map

The following is the strengthened continuity result for the mapping $\mathbf{VW}_f$ as discussed in §4.7.2.

Lemma 4.8.2. *The mapping $\mathbf{VW}_f$ is continuous, in its $L_2(G)$ dependence upon $\tilde{v}$ and $\tilde{w}$, and in its $H^1(G)$ dependence upon u, as a mapping into $\prod_1^2 H^1_\rho(G)$.*

Proof. We shall present the details completely in the case that lagging does occur in the gradient terms. Fundamental to the argument is an inequality which estimates the distance between two solution pairs in terms of preimages in the domain; later, one pair will be identified with a sequence member, and the other with the sequential limit. This, in a nutshell, is the crux of the sequential characterization of continuity. To proceed, one uses the sum of the representations (4.77, 4.78), and the elementary inequality,

$$|(f, g)| \le \frac{1}{2}\left(\eta^{-1}\|f\|^2 + \eta\|g\|^2\right), \tag{4.117}$$

for appropriate choices of η to obtain the following fundamental inequality:

$$\begin{aligned}
&\|v_1 - v_2\|^2_{H^1_\rho} + \|w_1 - w_2\|^2_{H^1_\rho} \qquad (4.118)\\
&\leq C[\|u_1 - u_2\|^2_{L_2} + \|\tilde{v}_1 - \tilde{v}_2\|^2_{L_2} + \|\tilde{w}_1 - \tilde{w}_2\|^2_{L_2} + \\
&\|\{|\exp(u_1 - \tilde{v}_1) - \exp(u_2 - \tilde{v}_2)| + \\
&|\exp(\tilde{w}_1 - u_1) - \exp(\tilde{w}_2 - u_2)|\}(|\nabla v_2| + |\nabla w_2|)\|^2_{L_2} + \\
&\|\{|\mu_n(\nabla u_1) - \mu_n(\nabla u_2)| + |\mu_p(\nabla u_1) - \mu_p(\nabla u_2)|\}(|\nabla v_2| + |\nabla w_2|)\|^2_{L_2}].
\end{aligned}$$

Here, C does not depend on the functions $\tilde{v}_i$ and $\tilde{w}_i$, $i = 1, 2$. To obtain (4.118), estimate the first terms on the right hand sides of (4.77) and (4.78) by (4.117) with $g = \sqrt{\mu_n}\nabla(v_1 - v_2)$, and $g = \sqrt{\mu_p}\nabla(w_1 - w_2)$, respectively. The choices of η are $\eta = \frac{\lambda_1 - \lambda_0}{2\lambda_1}e^{\gamma-\beta}$, $\eta = \frac{\lambda_1 - \lambda_0}{2\lambda_1}e^{\alpha-\delta}$, respectively. To estimate the second terms, the same choices of g and η are made. The estimation of $\frac{1}{2}\eta^{-1}\|f\|^2$ for the corresponding first terms in (4.77, 4.78) leads directly to terms dominated by a constant multiplier of the second to last term on the right hand side of (4.118). The estimation of $\frac{1}{2}\eta^{-1}\|f\|^2$ for the corresponding second terms leads to terms similarly dominated by the final term of (4.118). In order to estimate the recombination term differences, one proceeds from identities of the form,

$$\begin{aligned}
R_v(u_1, v_1, w_1) - R_v(u_2, v_2, w_2) &= [R_v(u_1, v_1, w_1) - R_v(u_2, v_1, w_1)] \\
&+ [R_v(u_2, v_1, w_1) - R_v(u_2, v_2, w_2)], \qquad (4.119)
\end{aligned}$$

with a similar identity involving R_w. The third terms on the right hand sides of (4.77) and (4.78) are handled by use of (4.119) and its R_w counterpart. The differences where the u argument is common, i.e., is equal to u_2, are estimated by use of the hypothesis (4.74). The remaining terms are separately estimated by (4.117), with $g = v_1 - v_2$, $g = w_1 - w_2$, respectively, and η given by $\eta = \frac{\lambda_0' - \lambda_0}{\lambda_1}e^{\gamma-\beta}$, $\eta = \frac{\lambda_0' - \lambda_0}{\lambda_1}e^{\alpha-\delta}$, respectively. Here, $\lambda_0 < \lambda_0' < \lambda_1$. For the estimation of $\eta^{-1}\|f\|^2$ in these cases, we use the assumed Lipschitz properties of R_v and R_w to obtain terms dominated by the first three terms on the right hand side of (4.118). With these choices, the left hand side of the sum of (4.77) and (4.78), after absorption of correlated right hand side $\|g\|^2$ terms, is seen to dominate

$$\left(\frac{1}{2} - \frac{\lambda_0'}{2\lambda_1}\right)\left(\exp(\gamma - \beta)\|v_1 - v_2\|^2_{H^1_\rho} + \exp(\alpha - \delta)\|w_1 - w_2\|^2_{H^1_\rho}\right),$$

whereas the $\|f\|^2$ terms of the right hand side of the sum are dominated by (4.118) for appropriate choices of C. This concludes the verification of (4.118). In order to use this inequality, we identify v_1 and w_1 with components v_k and w_k of a sequence of image points, under the map $\mathbf{VW}_f$, of a specified sequence of pairs in K, written $[\tilde{v}_k, \tilde{w}_k]$; the function u_k has the usual meaning of the image under $\mathbf{U}_f$. We further identify v_2 and w_2 with components v and w to be determined. The continuity property stated in the lemma amounts to the following:

– Suppose $[\tilde{v}, \tilde{w}]$ is the limit in $\prod_1^2 L_2(G)$ of the above sequence, and $u = \mathbf{U}_f([\tilde{v}, \tilde{w}])$. Then, for every subsequence $\{k_j\}$ of $\{k\}$, there is a further subsequence $\{k_{j_l}\}$ such that the image $[v_i, w_i]$ of $[\tilde{v}_i, \tilde{w}_i]$ converges to $[v, w]$, as i ranges over the subsequence $\{k_{j_l}\}$, in the product space $\prod_1^2 H_\rho^1(G)$.

The subsequence $\{k_{j_l}\}$ is determined as follows. For $i = k_{j_l}$:

1. $\tilde{v}_i$ converges pointwise to $\tilde{v}$, and $\tilde{w}_i$ converges pointwise to $\tilde{w}$;
2. u_i converges pointwise to u, and ∇u_i converges pointwise to ∇u.

This is possible by the result that L_2 convergence implies pointwise convergence of a subsequence. For the gradient terms, we have only local L_2 convergence, but a diagonalization argument still yields pointwise subsequential convergence. Given (1) and (2) above, we conclude the bounded pointwise convergence to zero of the multipliers of $|\nabla v_2| + |\nabla w_2|$ on the right hand side of (4.118). One concludes from this, via the Lebesgue dominated convergence theorem, that the integrals representing those terms tend to zero also. The first three terms on the right hand side trivially tend to zero. This completes the proof of continuity in the case of lagging in the gradient multiplier. The case of no lag is even simpler, and is based upon the representations (4.79) and (4.80). We omit the details.

Remark 4.8.1. An enhanced result of this type is particularly interesting when one is establishing the foundations of an operator calculus. For example, in the following chapter, in Lemma 5.4.3, an even stronger result is established in the special case when the recombination term is zero, and the mobilities are constant. In such a simpler structural setting, the mapping is even Lipschitz continuous, and $\mathbf{VW}_f$ decomposes nicely into $[\mathbf{V}_f, \mathbf{W}_f]$ in this case, of course. In this chapter, the nonvanishing of R is responsible for much of the structure of the decoupling mappings. The generally accepted form of R, in the absence of avalanche generation, is given by the equation (4.70):

$$R(u, v, w) = F(u, v, w)(\exp(w - v) - 1),$$

where F is the sum of its Shockley-Read-Hall and Auger multipliers, respectively:

$$\begin{aligned} F(u, v, w) &= \frac{c_1}{1 + c_2 \exp(u - v) + c_3 \exp(w - u)} \\ &+ (c_4 \exp(u - v) + c_5 \exp(w - u)). \end{aligned} \tag{4.120}$$

Here, the (physical) constants are positive:

$$c_i > 0, \ i = 1, 2, 3, 4, 5.$$

Remark 4.8.2. We note that when Auger recombination is suppressed, i.e., when $c_4 = c_5 = 0$, then R is monotone increasing in w and monotone decreasing in v. In this special case, complete decoupling of the system is achieved in an admissible manner by lagging w completely in the v-equation, and by lagging v completely in the w-equation. Otherwise, if Auger recombination is present, as has been remarked earlier, a lagging guaranteed to be admissible is achieved by lagging F completely, and keeping the remaining factor in (4.70) unlagged in both equations of the coupled current continuity subsystem.

Avalanche generation, on the other hand, entails complicated gradient expressions. Nonetheless, the maximum principle arguments carried out in [67] for v and w are still valid when this term is added to the recombination, because of the sign of the avalanche term (see [100] for the precise expressions). Other parts of the analysis there are no longer valid, however, in this more general situation. Note that the unbounded gradients, due to boundary transition points, present a technical difficulty.

5. Nonlinear Convergence Theory for Finite Elements

In the following chapter (Chap. 6), a very powerful convergence theory will be established for numerical fixed point approximations. It is the goal of the present chapter to show that this framework encompasses the semiconductor device model in the case of constant mobility coefficients and zero recombination terms. The system can then be written, with suitable scaling, as,

$$-\nabla\cdot(\epsilon\nabla u)+e^{u-v}-e^{w-u} = k_1, \quad (5.1)$$

$$-\nabla\cdot(e^{u}\nabla e^{-v}) = 0, \quad (5.2)$$

$$-\nabla\cdot(e^{-u}\nabla e^{w}) = 0, \quad (5.3)$$

subject to mixed Dirichlet/homogeneous Neumann boundary conditions, taken on the Dirichlet part, Σ_D, of the device boundary, and on its complement, Σ_N, respectively. These are specified on Σ_D by the traces of appropriate functions $\bar{u}$, $\bar{v}$, $\bar{w}$, where $\bar{u}$, $\bar{v}$, $\bar{w}$ are in $C^2(\bar{G})$, and specified on Σ_N weakly. The analysis to follow admits generalization to spatially dependent, positive diffusion and mobility coefficients. It remains to be determined what extensions beyond this are possible. The existence and maximum principle analysis for the above boundary-value problem, and extensions thereof, have been derived in Chap. 4. Throughout this chapter, we shall assume that G is a polyhedral domain. The final nonlinear Galerkin finite element approximation results, embodied in Corollary 5.8.1, are valid in Euclidean space of dimension $N \leq 3$, but we shall present the preliminary results more generally, and indicate stipulations regarding N when appropriate.

The chapter contains an assumption dealing with the triangulation, and four additional assumptions, all dealing with the regularity of solutions, or stability of approximations. Assumption 5.1 operates throughout, while the remaining assumptions are required only for a selection of the preliminary results. All are required for Lemmas 5.7.1 and 5.7.2, which lead directly to Corollary 5.8.1. The actual regularity of the solution of the mixed boundary value problem with homogeneous conditions on the Dirichlet boundary has been presented in [106]. The conversion to inhomogeneous conditions has been worked out by Kerkhoven in [82]. In two dimensions, the gradient regularity in the inhomogeneous case is L_p, $p < 4$, while the weaker L_3 regularity holds in three dimensions for the most general singularities. In fact, the authors of [106] consider regularity in three sets, near Dirichlet, Neumann, and transition points, respectively. It is the Neumann points which limit the gradient regularity to L_3 for $N = 3$; $p < 4$ holds near transition points. The assumptions of this chapter deal with this issue, either directly or indirectly. Assumption 5.3, for example, assumes that the three-dimensional singularities actually present are not the most general possible. This is consistent with the manner in which semiconductor boundary conditions are imposed with respect to the Dirichlet boundary. We shall now develop the system mappings required for the results. A sketch of this theory, without detailed proofs, and based upon somewhat different ideas, first appeared in [74]. Although the statements of the results of this chapter deal exclusively with the convergence of Galerkin approximations based on piecewise linear subspaces, the underlying machinery, when coupled with the linearization results of the next chapter, provides a basis for reducing approximations to linearization procedures based on the Gummel map. This is detailed in the final chapter (see especially §7.1).

An important comment relates to the choice of discretization. It is known that exponential fitting methods are essential to the proper discretization of drift-diffusion systems (cf. [116]). It is also known that, when the quasi-Fermi levels are employed, appropriate quadrature formulas for the integrals resulting from piecewise linear finite element procedures yield the proper exponential fit (cf. [131]). It is for this reason that the analysis of piecewise linear finite elements is important and decisive.

5.1 Definitions of the Composite Mappings of T

We provide definitions of **T** and associated composition mappings, as defined on open sets in function space. In Chap. 4 and elsewhere, it suffices to define the mappings on closed convex sets, specified by the maximum principles, It follows from the theory of Chap. 4 that a fixed point mapping may be defined to act invariantly on the set K, defined by (4.51). In the setting of this chapter, the maximum principles may be strengthened, due to the absence of the recombination term. Thus, K may be defined here as

$$K = \{[v, w] : \alpha_v \leq v \leq \beta_v,\ \alpha_w \leq w \leq \beta_w\}, \tag{5.4}$$

where

$$\alpha_v = \inf_{\Sigma_D} \bar{v}, \qquad \alpha_w = \inf_{\Sigma_D} \bar{w}, \tag{5.5}$$

$$\beta_v = \sup_{\Sigma_D} \bar{v}, \qquad \beta_w = \sup_{\Sigma_D} \bar{w}. \tag{5.6}$$

Because this set, defined by the maximum principles for the quasi-Fermi levels, is not open, and because the differential operator calculus requires the extension to sets with interior, we construct the definition of **T** such that the assumption, that the preimage $[v, w]$ lies in K, can be removed. To achieve this, we employ a preconditioning truncation operator **Tr**, which leaves $[v, w]$ unaffected within K (where the solution lies), but which restricts the range to a set K_1 (cf. (5.8)), which is only slightly larger than K. By carefully selecting K_1, we achieve the result that the intermediate function $u = U$ in the definition of **T** satisfies 'a priori' L_∞ bounds, which are only slightly more relaxed than those for u in (4.56) and (4.57). These bounds, expressed in (5.10) to follow, are valid as the solution U of (5.1) takes on the range of a map defined for all $[v, w]$ *in an appropriate open subset of* $\prod_1^2 H^1$, and not simply on the set K. This open subset is denoted Ω below.

Definition 5.1.1. *We introduce* $h_i \in C_0^\infty(\mathbb{R})$, $i = 1, 2$, *such that the support of* h_i *is the interval,* $[\alpha_i, \beta_i]$, $i = 1, 2$, *and*

$$\begin{aligned} h_1(t) &= t, \ \alpha_v \le t \le \beta_v, \\ h_2(t) &= t, \ \alpha_w \le t \le \beta_w. \end{aligned}$$

Below we shall define an open ball Ω, *centered at zero, in* $\prod_1^2 H^1$, *on which*

$$\mathbf{Tr}[v, w] := [\mathbf{h}_1(v), \mathbf{h}_2(w)], \qquad [v, w] \in \Omega. \tag{5.7}$$

Note that the range of **Tr** *is contained in* $K_1 \subset \prod_1^2 L_\infty$, *where*

$$K_1 = \{[v, w] \in \prod_1^2 L_\infty : \alpha_1 \le v \le \beta_1, \ \ \alpha_2 \le w \le \beta_2\}. \tag{5.8}$$

We consider maps **U**, **V**, *and* **W** *with elements in the domain of* **U** *taken from a subset of* K_1. *In terms of these operators,* **T** *may be defined by*

$$\mathbf{T} = [\mathbf{V} \circ \mathbf{U} \circ \mathbf{Tr}, \mathbf{W} \circ \mathbf{U} \circ \mathbf{Tr}]. \tag{5.9}$$

It is essential to identify carefully the domains and ranges of the composition maps used to define **T**. Since

$$|\nabla[\mathbf{h}_1(v)]|^2 = |\mathbf{h}_1'(v)\nabla v|^2 \le c|\nabla v|^2,$$

with a similar inequality for $|\nabla[\mathbf{h}_2(w)]|^2$, it follows that the mapping **Tr** has range contained in the set $K_1 \cap \{C\mu : \mu \in \Omega \subset \prod_1^2 H^1\}$, for some positive constant C, which depends upon the radius of Ω by the gradient inequalities just cited. Thus, we select the domain of **U** to be $K_1 \cap (C\Omega)$.

Remark 5.1.1 (Γ as a superset for the range of **U***).* Employing the H^1 norm defined below in (5.14), we see that the range of **U** is contained in a bounded set Γ in $H^1(G) \cap L_\infty(G)$; indeed, the pointwise bounds, involved in defining Γ, follow from arguments paralleling the derivation of (4.56) and (4.57), while the H^1 bounds for **U** and its finite element approximation, also involved in defining Γ, will be elaborated in §5.2.3 (cf. Lemma 5.2.1). In particular, the following pointwise bounds (maximum principles) hold for $U = \mathbf{U}[v, w]$:

$$\gamma \leq U \leq \delta,$$

$$\gamma = \min(\gamma', \inf_{\Sigma_D} \bar{u}), \qquad \delta = \max(\delta', \sup_{\Sigma_D} \bar{u}),$$

$$\gamma' = \sinh^{-1}[(1/2) \inf_G k_1 \, e^{(\alpha_1 - \alpha_2)/2}] + (\alpha_1 + \alpha_2)/2, \tag{5.10}$$

$$\delta' = \sinh^{-1}[(1/2) \sup_G k_1 \, e^{(\beta_1 - \beta_2)/2}] + (\beta_1 + \beta_2)/2.$$

Definition 5.1.2. *Mappings* $\mathbf{V}$ *and* $\mathbf{W}$, *whose evaluations,*

$$v = \mathbf{V}(U), w = \mathbf{W}(U),$$

are quasi-Fermi levels, are naturally defined as in earlier chapters, by solution of the decoupled equations (5.2, 5.3), *with* $u = U$*:*

$$-\nabla\cdot (e^U \nabla e^{-v}) = 0, \tag{5.11}$$

$$-\nabla\cdot (e^{-U} \nabla e^{w}) = 0, \tag{5.12}$$

Finally, the joint domain of $\mathbf{V}$ *and* $\mathbf{W}$ *is* Γ, *while the range of these maps is contained in the intersection of* K *(not* K_1*), with the domain* Ω, *which is now discussed.*

Since, for $U \in \Gamma$ (the pointwise bounds suffice),

$$\begin{aligned} \int_G |\nabla v|^2 dx &\leq \exp(\beta_v - \gamma) \int_G e^{U-v} |\nabla v|^2 dx \\ &= \exp(\beta_v - \gamma) \int_G e^{U-v} \nabla v \cdot \nabla \bar{v} dx \qquad (5.13) \\ &\leq (1/2)[\int_G |\nabla v|^2 dx + \exp(2(\delta + \beta_v - \gamma - \alpha_v)) \int_G |\nabla \bar{v}|^2 dx], \end{aligned}$$

it follows, for $\|\cdot\|_{H^1}^2$ given by

$$\|v\|_{H^1}^2 = \|\nabla v\|_{L_2}^2 + (\int_{\Sigma_D} v dx)^2, \tag{5.14}$$

that

$$\|v\|_{H^1}^2 < e^{2(\delta - \gamma + (\beta_v + \beta_w) - (\alpha_v + \alpha_w))} \|\bar{v}\|_{H^1}^2 \quad (\text{Note: } v\,|_{\Sigma_D} = \bar{v}\,|_{\Sigma_D}),$$

with a similar estimate for $\|w\|_{H^1}^2$. Thus, we initially choose Ω to contain the open ball centered at 0 of radius,

$$\rho = e^{(\delta - \gamma + (\beta_v + \beta_w) - (\alpha_v + \alpha_w))} \|[\bar{v}, \bar{w}]\|_{\prod H^1}.$$

It is evident that Ω contains the range of a reduced fixed point map $\mathbf{T}_0$, with domain K, as well as that of $\mathbf{T}$, and hence, that $\mathbf{T}$ is a proper extension of $\mathbf{T}_0 |_{\mathbf{Range}\, \mathbf{T}_0}$. In particular, $\mathbf{T}$ possesses a fixed point in Ω by the results of Chap. 4. However, it is essential for our purposes that Ω also contain the range of $\mathbf{T}_n$, to be defined shortly (cf. (5.23)) in §5.2.3. The composite mappings, $\mathbf{U}_h$, $\mathbf{V}_h$, and $\mathbf{W}_h$, will be defined there also. Pointwise estimates for these mappings parallel those for the continuous problem under mesh restrictions, and are summarized in §5.2.2. Energy estimation of $v_h = \mathbf{V}_h(U_h)$ and $w_h = \mathbf{W}_h(U_h)$ is required to complete the argument. An adjustment of (5.13), in which $\bar{v}$ is replaced by $\bar{v}_I$ and v is replaced by v_h, yields the result that the number ρ just defined need only be perturbed by a term of order $O(h)$. To obtain an estimate for this perturbation, we may estimate $\|\nabla(\bar{v} - \bar{v}_I)\|_{L_2}$ by the stronger inequality used in (5.29). This gives us, finally, an admissible radius of Ω.

Remark 5.1.2 (truncation map). A serious technical problem, which will be dealt with fully later in this chapter, is the lack, in the energy norm, of Lipschitz continuity of the derivative of the mapping, $\mathbf{Tr}$. This necessitates, in the application of the convergence theory of the following chapter, the introduction of a stronger norm than the energy norm alone, viz., one involving the minimum of the energy norm and the L_∞ norm. It will happen that the derivative of $\mathbf{Tr}$ is Lipschitz continuous on the intersection of Ω with an L_∞ hyperplane, in the joint topology. This explains why we develop, in the following chapter, a generalized version of the Krasnosel'skii operator calculus, admitting this more general situation. The strength of the convergence theory is that, although certain pointwise estimates are required to tend to zero, this need not occur at the same rate as the energy estimates.

Throughout this chapter, we shall employ a regularization hypothesis for $\mathbf{T}$. It may be reduced to the following statement, though there is at least an implicit coupling to Assumption 5.3 to follow in §5.3.3.

Assumption 5.1 (fundamental regularization hypothesis). The solution maps $\mathbf{V}$, $\mathbf{W}$ are regularizing, i.e., they map Γ boundedly into $H^{1+\theta}(G)$ for some $\theta > 0$. Here, Γ is introduced in Remark 5.1.1.

Only in Assumption 5.3 do we make this hypothesis more specific in terms of linkages between θ and N. In general, it is expected that θ varies with Euclidean dimension, in a manner consistent, via the Sobolev embedding theorem, with the gradient regularity known for the mixed boundary-value problem. In particular, for $N = 3$, θ would be expected to be at least $1/2$, while for $N = 2$, the regularization would be expected to hold for all $\theta < 1/2$. However, at this point, we assume only $\theta > 0$.

Remark 5.1.3. We close this section by noting that the maximum principles (5.4) and (5.10) are special cases of the (reduced) bounds,

$$\gamma \le u \le \delta, \tag{5.15}$$

discussed in [84], where γ, δ have the meaning of (5.10), but

$$\gamma' = \inf_{x \in G} f^{-1}(x, \inf_{y \in G} g(y)), \qquad \delta' = \sup_{x \in G} f^{-1}(x, \sup_{y \in G} g(y)), \tag{5.16}$$

for the solution u of the gradient equation,

$$-\nabla \cdot [a(x)\nabla u(x)] + f(x, u(x)) = g(x). \tag{5.17}$$

Here, $a \geq a_0 > 0$ with $a, g \in L_\infty$, and f is increasing and locally Lipschitz in u for each $x \in G$, with $f^{-1}(x, \cdot)$ the corresponding inverse. For (5.10), f is given by (5.26).

5.2 The Numerical Map $\mathbf{T}_n$

In this section, we introduce the piecewise linear finite element method which defines the numerical fixed point map. We also describe the associated approximation properties. We assume a given family of simplicial decompositions of the polyhedral domain G (assumed polyhedral in this chapter only) with the property of quasi-uniformity (cf. [64]). With respect to actual computational implementation of these finite element approximations, appropriate upwinding for the current continuity equations is achieved by quadrature rules which are more suitable than exact evaluation. This is related to the exponential fitting discussion of Chap. 2. Our focus here is on the approximation question rather than the details of the discretization, per se.

5.2.1 The Composite Finite Element Maps

Here we introduce the relevant composite finite element mappings. For reasons of strict symmetry, the domain of $\mathbf{U}_h$ is chosen to be the same set $K_1 \cap (C\Omega)$ as the domain of $\mathbf{U}$. The finite element equations for the *uncoupled* Poisson equation are given in duality pairing notation by,

$$\langle \epsilon(x)\nabla U_h, \nabla \phi_i \rangle + \langle e^{U_h - v} - e^{w - U_h}, \phi_i \rangle - \langle k_1, \phi_i \rangle = 0 \quad \text{for } i = 1, \cdots, M, \tag{5.18}$$

where $U_h = \mathbf{U}_h[v, w]$ is a finite element function, and the ϕ_i are appropriate test functions comprising a nodal basis of the piecewise linear finite element subspace S_h. According to the assumptions specified regarding $\bar{u}$, it follows that we may select the piecewise linear interpolant $\bar{u}_I$ of $\bar{u}$ so that $U_h \in \bar{u}_I + S_h$, where the members of S_h vanish on the Dirichlet boundary Σ_D of the polyhedral domain G. The functions of S_h are continuous and are linear in each simplex, S.

Next, we characterize the terms $\mathbf{V}_h(U_h) = v_h$, $\mathbf{W}_h(U_h) = w_h$, by

$$\langle e^{U_h - v_h} \nabla v_h, \nabla \phi_i \rangle = 0, \quad \text{for } i = 1, \cdots, M, \tag{5.19}$$

$$\langle e^{w_h - U_h} \nabla w_h, \nabla \phi_i \rangle = 0, \quad \text{for } i = 1, \cdots, M. \tag{5.20}$$

$\mathbf{V}_h$ and $\mathbf{W}_h$ are actually defined on the more inclusive set Γ in the obvious way; energy and pointwise estimates are unaltered. As before, $\phi_i \in S_h$, and $v_h \in \bar{v}_I + S_h$, $w_h \in \bar{w}_I + S_h$, where $\bar{v}_I$ and $\bar{w}_I$ are interpolants of $\bar{v}$ and $\bar{w}$, respectively.

The range of $\mathbf{U}_h$ is contained in Γ (as is the range of $\mathbf{U}$), which serves as the joint domain of $\mathbf{V}_h$ and $\mathbf{W}_h$. Indeed, the fact that U_h satisfies the bounds (5.10) will be discussed in the following subsection. A similar statement applies to the discrete bounds satisfied by v_h and w_h. The only unverified issue related to the definition of $\mathbf{T}$ then is the derivation of the H^1 bounds associated with Γ, containing the range of $\mathbf{U}$. This matter, and the issues related to the range of $\mathbf{U}_h$ within Γ, as well as the existence questions for the finite element solutions, will be taken up in §5.2.3.

5.2.2 The Discrete Maximum Principles

Consider the gradient equation, (5.17). We shall describe conditions on the simplicial decomposition for the discrete maximum principles to hold for the mixed boundary-value problem.

Definition 5.2.1 (simplicial decomposition). *Let S be an N-dimensional simplicial finite element such that*

1. *V is the volume;*
2. *$\mathbf{v}_i$ is a vertex;*
3. *e_{ij} is the edge connecting vertices $\mathbf{v}_i$ and $\mathbf{v}_j$;*
4. *F_k is the face opposite the k-th vertex, with measure $|F_k|$;*
5. *h_i is the normal distance of $\mathbf{v}_i$ to F_i;*
6. *γ_{ij} is the angle between the inward normal vectors to the faces F_i and F_j;*
7. *ϕ_l is the piecewise linear nodal basis function which is 1 at vertex $\mathbf{v}_l$;*
8. $$\alpha_{ij} \equiv \int_S a(x) \nabla\phi_i \cdot \nabla\phi_j \, dx$$
is the ijth entry of the element stiffness matrix;
9. *$\langle a(x)\rangle \equiv \int_S a(x)dx/V$ is the average of $a(x)$ over the element S;*
10. *a_{ij} is the ijth element of the assembled stiffness matrix.*

Remark 5.2.1. It was shown in [84, Lemma A.1] that

$$\alpha_{ij} \equiv \int_S a(x)\nabla\phi_i \cdot \nabla\phi_j dx = \langle a(x)\rangle \cos(\gamma_{ij}) \frac{1}{h_i h_j} V, \tag{5.21}$$

or

$$\alpha_{ij} = \langle a(x)\rangle \cos(\gamma_{ij}) \frac{|F_i||F_j|}{N^2 V}.$$

In [84, Theorem 3.2] it was also shown that approximate solutions of the gradient equation (5.17) satisfy discrete maximum principles, defined by the bounds, (5.15, 5.16). Hypotheses implying these bounds are given in the following.

Assumption 5.2 (triangulation and Lipschitz hypotheses). a) In dimension $N \geq 2$, we require that, for every edge e_{jk}, the off-diagonal element a_{jk} in the stiffness matrix satisfies,

$$a_{jk} = \sum_{S \text{ adjacent } e_{jk}} \langle a(x)\rangle_S \cos(\gamma_{jk}^{(S)}) \frac{V^{(S)}}{h_j h_k} \leq -\frac{\rho}{h_{\max}^2} \sum_{S \text{ adjacent } e_{jk}} V^{(S)},$$

with $\rho > 0$.
In two dimensions, the well-known requirement that, for every edge e_{jk} in the triangulation, we have the inequality,

$$\tfrac{1}{2}[\langle a(x)\rangle_{T_1} \cot(\omega_1) + \langle a(x)\rangle_{T_2} \cot(\omega_2)] \geq \rho > 0,$$

where the T_i are the two triangles adjacent to edge e_{jk}, and the ω_i are the two angles opposite to the edge e_{jk}, is a slightly more restrictive version of this condition. In higher dimensions, the above hypothesis generalizes in a significant way the condition derived by the authors of [29]. These authors imposed the sufficient condition that the angle between the vectors normal to any two faces of the same polyhedron in the mesh has to be bounded uniformly from above by $\pi/2 - \eta$.

b) For all $c, d \in \mathbb{R}, c < d : |f(x,u) - f(x,v)|/|u - v| \leq D(d,c)$ if $c \leq u, v \leq d$, where $D(\cdot,\cdot)$ is a Lipschitz constant which is a monotonically increasing function of d and a monotonically decreasing function of c.

c) The numbers h_i satisfy $h_i \geq h_0 h$, where h_0 does not depend on h.

d) There is a (stringent) condition that h must be sufficiently small. This is expressed in terms of a uniform Lipschitz constant D for f in its second argument, in terms of ρ, and in terms of Euclidean dimension N:

$$h_{\max}^2 \leq \rho \frac{(N+1)(N+2)}{D}. \tag{5.22}$$

The proofs of these results may be found in [84], where full details are given. The mesh condition given in (5.22) arises from the equilibration of gradient and undifferentiated terms. Specifically, the term (5.21) dominates locally the term computed from

$$\int_S \phi_i \phi_j \, dx = \frac{V}{(N+1)(N+2)},$$

where $i \neq j$ (This formula is derived in the proof of [84, Theorem 3.2]). Although these ideas are closely related to the theory of M-matrices (cf. [134]) and M-functions (cf. [110]), they are not directly deducible from these theories.

5.2.3 The Numerical Fixed Point Map

We assume the hypotheses of the previous subsection, so that the discrete maximum principles hold for the component mappings. If the given range space for the joint mapping, $[\mathbf{V}_h, \mathbf{W}_h]$, has dimension n, then we may define $\mathbf{T}_n$, the numerical fixed point map, by

$$\mathbf{T}_n = [\mathbf{V}_h \circ \mathbf{U}_h \circ \mathbf{Tr}, \mathbf{W}_h \circ \mathbf{U}_h \circ \mathbf{Tr}]. \tag{5.23}$$

By viewing both the solution of the Poisson equation and the finite element equation as resulting from convex minimization, one obtains H^1 bounds common to both, and hence the completion of the definition of Γ (A more precise definition of the range space is given in §5.4).

Thus, define the convex functional,

$$\Phi(u) = \frac{1}{2}\int_G \varepsilon|\nabla u|^2 + \int_G H(\cdot, u) - \int_G (k_1 - f(\cdot, 0))u, \tag{5.24}$$

where

$$H(x,t) = \int_0^t [f(x,s) - f(x,0)]\, ds \geq 0, \tag{5.25}$$

$$f(x,s) = e^{s-v(x)} - e^{w(x)-s}, \tag{5.26}$$

for a given pair v, w. U and U_h minimize this functional over $\bar{u} + H^1_{0,\Sigma_D}$ and $\bar{u}_I + S_h$, respectively. Here, H^1_{0,Σ_D} is the completion, as in Chap. 4, of $C_0^\infty(G \cup \Sigma_N)$ in the norm defined by (5.14) (equivalently, zero trace on Σ_D), and $S_h \subset H^1_{0,\Sigma_D}$. In particular,

$$\Phi(U) \leq \Phi(\bar{u}), \qquad \Phi(U_h) \leq \Phi(\bar{u}_I). \tag{5.27}$$

H^1 estimates, and hence a proper bound for the definition of Γ, are obtained for $u = U$ and $u = U_h$ from

$$\frac{1}{2}\inf \epsilon \int_G |\nabla u|^2 \leq \Phi(u) - \int_G H(\cdot, u) + \int_G (k_1 - f(\cdot, 0))u$$

$$\leq \max(\Phi(\bar{u}), \Phi(\bar{u}_I)) + \frac{1}{2}\int_G |k_1 - f(\cdot, 0)|^2 + \frac{1}{2}\int_G |\, u\,|^2 . \tag{5.28}$$

The second and third terms are estimated from the maximum principles, and the assumption $k_1 \in L_\infty(G)$. Now $\Phi(\bar{u}_I) \leq \Phi_0$, independent of h, a fact which follows from (cf. [80, p. 85] and [127]):

$$\begin{aligned} \|\nabla(\bar{u}_I)\|_{L_\infty} &\leq \|\nabla(\bar{u} - \bar{u}_I)\|_{L_\infty} + \|\nabla \bar{u}\|_{L_\infty} \\ &\leq \frac{Ch^2}{\min_{i=1,\cdots,M} h_i}|\bar{u}|_{W^{2,\infty}} + \|\nabla \bar{u}\|_{L_\infty}, \end{aligned} \tag{5.29}$$

and the quasi-uniformity of the mesh, whereby $\min h_i \geq h_0 h$. It follows that gradient estimates, and hence H^1 estimates, are uniformly obtainable for the continuous and discrete problems. The existence of solutions of the finite element equations (5.19, 5.20) is demonstrated by convex minimization, as for (5.18). This completes the discussion concerning the definition of $\mathbf{T}_n$ and $\mathbf{T}$. We present the key conclusion as follows.

Lemma 5.2.1. *The solutions U and U_h, derived from the minimization of (5.24), have H^1 norms uniformly less than a fixed positive constant, depending only on the prescribed data and on G.*

5.3 Approximation Theory in Energy Norms and Pointwise Norms

Prior to describing the approximation properties of $\mathbf{T}_n$, it is essential to discuss the linear approximation properties of the H^1_{0,Σ_D} projection $\mathbf{Q}_h$ onto S_h. For $H^{1+\theta}(G) \cap H^1_{0,\Sigma_D}(G)$ functions, with norm bounded uniformly in this space, interpolation space theory (cf. [14]) gives a uniform h^θ estimate, as measured in H^1, provided the approximation process is capable of yielding an $O(h)$ estimate for $H^2(G) \bigcap H^1_{0,\Sigma_D}(G)$. Although detailed results for the latter estimate are not available in the literature, the general procedure is described adequately in [128] for $N \leq 3$, by a direct argument applied to the piecewise linear interpolant taken over $\bar{G}$. Although our final approximation results are restricted to this case, we comment on the general case for completeness. The reader who wishes to skip this somewhat technical discussion can proceed directly to §5.3.1.

General N is considered in [127]: the piecewise linear interpolant of an extension/smoothing process gives the requisite energy upper bound, but the smoothing should be done only in the tangential variables on Σ_D, so that the smoothed function also vanishes on Σ_D. This can be made precise by a decomposition of the polygonal boundary of G, according to edges and internal boundaries of Σ_D, with an accompanying finite partition of unity for G (see [62, §4] for an illustration of how this can be done). In fact, the partition of unity is for an open set containing G, and those sets of the partition which intersect ∂G are assumed to do so in components of Σ_D and Σ_N. The partition of unity sets are then translated and rotated into canonical position. The extension process uses a variant of the Calderón extension operator, applied to cylindrical translation/rotation domains and canonical variables, for Σ_N component "cylindrical" cross sections. For Σ_D component cross sections, the extension is the trivial zero extension. The entire extended function may be reassembled, via the partition of unity. The smoothing can also be carried out in the canonical variables if desired. When the errors, either in the smoothing or in the polynomial interpolation, are estimated, it is advantageous to consider each partition of unity domain in G separately. In particular, in the case of polynomial interpolation estimation, one must consider each partition set intersecting a given simplex over the triangulation. For readers familiar with the constructions of [127], the Fourier transform is applied only in the tangential variables, for partition of unity domains with Σ_D cross sections, to estimate the smoothing error, via Strang's argument. In estimating the piecewise linear interpolation error relative to the smoothing, we note that interpolation is *also* specified on Σ_N. It is assumed, in the sense that the arguments require it, that the topology of Σ_D and Σ_N allows for the mechanics sketched above for the extension. This completes the $O(h)$ estimate, and interpolation space theory now applies. Improved estimates, based upon graded meshes as currently employed in the h-p theory, tend not to be uniform over Sobolev classes, and do not appear to be applicable here. We are not aware either, of maximum principles for degree $p > 1$.

5.3.1 Approximation Theory for Gradient Equations

The next result is a generic result for gradient equations which will be used to deduce the approximation properties of U_h. The symbol, $\langle \cdot \rangle$, has its usual meaning as the duality pairing.

Lemma 5.3.1. *Suppose $a(\cdot,\cdot)$ is a continuous symmetric bilinear form on H^1, which is L_2 coercive on H^1_{0,Σ_D}. For $u \in H^1$, let $F(u)$ denote the continuous linear functional on H^1_{0,Σ_D}, defined by*

$$F(u)(z) = \int_G f(\cdot, u)z, \tag{5.30}$$

for f increasing in its second argument and $\partial f(\cdot, s)/\partial s \leq C$. Suppose that u and u_h satisfy the gradient relations,

$$\begin{aligned} a(u,z) + F(u)(z) &= \langle g, z\rangle, \quad \forall z \in H^1_{0,\Sigma_D}, &(5.31)\\ a(u_h,z_h) + F(u_h)(z_h) &= \langle g, z_h\rangle, \quad \forall z \in S_h, &(5.32)\end{aligned}$$

where $u \in \bar{u} + H^1_{0,\Sigma_D}, u_h \in \bar{u}_I + S_h, \bar{u} \in C^2(\bar{G})$. Here, $g \in L_2$ is prescribed. Then there exist constants C_1 and C_2, independent of h, such that

$$a(u-u_h, u-u_h) \leq C_1 \inf_{z_h \in S_h} a(u-\bar{u}_I - z_h, u-\bar{u}_I - z_h) + C_2\|\bar{u}-\bar{u}_I\|^2_{H^1}. \tag{5.33}$$

Finally, there is no loss of generality in assuming that $u - \bar{u}_I$, on the right hand side of (5.33)*, vanishes on the Dirichlet boundary.*

Proof. The first step is the reduction to the case $\bar{u} \mapsto \bar{u}_I$. By this is meant that there is no loss of generality in assuming that the boundary data function in (5.31) may be taken to be $\bar{u}_I$. Denoting the solutions of the corresponding gradient equations with boundary data $\bar{u}, \bar{u}_I$, respectively, by

$$u_\phi = \phi + \bar{u}, \quad u_\psi = \psi + \bar{u}_I, \quad \phi, \psi \in H^1_{0,\Sigma_D}(G), \tag{5.34}$$

we obtain, upon subtraction and upon setting $z = \phi - \psi$,

$$\begin{aligned} a(\phi - \psi, \phi - \psi) &+ [F(u_\phi) - F(u_\psi)](u_\phi - u_\psi) = \\ &- a(\bar{u} - \bar{u}_I, \phi - \psi) + [F(u_\phi) - F(u_\phi)](\bar{u} - \bar{u}_I). \end{aligned}$$

By the continuity and L_2-coerciveness properties of $a(\cdot, \cdot)$, and the monotonicity and derivative properties of f, we deduce that

$$a(\phi - \psi, \phi - \psi) \leq c\|\bar{u} - \bar{u}_I\|_{H^1}^2, \tag{5.35}$$

for some constant c, which depends only on the bound for $\partial f / \partial s$ and the L_2 coerciveness constant of $a(\cdot, \cdot)$.

The second step is the consideration of (5.31, 5.32), with $\bar{u}$ replaced by $\bar{u}_I$ in (5.31). We shall show that (note that u_ψ above is here identified as u)

$$a(u - u_h, u - u_h) \leq c_1 \inf_{z_h \in S_h} a(u - \bar{u}_I - z_h, u - \bar{u}_I - z_h), \tag{5.36}$$

which, with (5.35), will yield (5.33). We begin with

$$\begin{aligned} & \tfrac{1}{2}a(u - u_h - z_h, u - u_h - z_h) \\ + \; & F(u)(u - u_h - z_h) - F(u_h)(u - u_h - z_h) \\ = \; & \tfrac{1}{2}a(u - u_h, u - u_h) + [F(u) - F(u_h)](u - u_h) + \tfrac{1}{2}a(z_h, z_h) \\ - \; & a(u - u_h, z_h) - [F(u) - F(u_h)](z_h) \\ \geq \; & \tfrac{1}{2}a(u - u_h, u - u_h), \end{aligned}$$

which follows from the weak relations (5.31, 5.32), the assumed monotonicity of f, and the nonnegativity of $a(z_h, z_h)$. It follows that, for arbitrary $\delta > 0$,

$$\begin{aligned} a(u - u_h, u - u_h) \; &\leq \; a(u - u_h - z_h, u - u_h - z_h) \\ &+ \; \delta C^2\|u - u_h\|_{L_2}^2 + \delta^{-1}\|u - u_h - z_h\|_{L^2}^2, \end{aligned}$$

when C bounds $\partial f / \partial s$. Since $u - u_h$ vanishes on Σ_D, we may choose δ so that (via L_2 coerciveness of $a(\cdot, \cdot)$)

$$\delta C^2\|u - u_h\|_{L^2}^2 \leq \tfrac{1}{2}a(u - u_h, u - u_h).$$

Since $u - u_h - z_h$ vanishes on Σ_D, it follows that

$$a(u - u_h, u - u_h) \leq c_1 a(u - u_h - z_h, u - u_h - z_h)$$

for all $z_h \in S_h$. The identification, $u_h + z_h \mapsto z_h + \bar{u}_I$, now yields (5.36).

Remark 5.3.1. The next result will be used to deduce the approximation properties of v_h and w_h. The proof operates by a series of reductions. The first involves the replacement of boundary data by interpolation in the equations defining v and w. This is followed by replacement of the arguments in the exponential functions within the divergence structure of the same equations. The function v_h (resp. w_h) can be defined as the finite element approximation, of the solution of the final reduced problem in the proof, via a *linear* projection operator, onto (linear) $Sp\{\bar{v}_I, S_h\}$ (resp. $Sp\{\bar{w}_I, S_h\}$).

Lemma 5.3.2. *The solutions of* (5.19, 5.20) *satisfy the inequality, for* $N \leq 3$ *and under Assumption* 5.3 *of* §5.3.2 *to follow,*

$$\begin{aligned} \|v - v_h\|^2_{H^1} + \|w - w_h\|^2_{H^1} &\leq C[\|U - U_h\|^2_{H^1} + \|\bar{v} - \bar{v}_I\|^2_{H^1} \\ + \|v^*_h - \mathbf{P}^1_n v^*_h\|^2_{H^1} + \|\bar{w} - \bar{w}_I\|^2_{H^1} &+ \|w^*_h - \mathbf{P}^2_n w^*_h\|^2_{H^1}], \end{aligned} \tag{5.37}$$

where U, v, *and* w *retain their meaning as images of composition maps (hence, solutions of gradient equations), and* C *is invariant over the domain of these maps. The function* v^*_h *(analogously,* w^*_h*) is described in the proof below as a solution of a linear equation, with interpolated boundary data* (cf. (5.41))*, and* $\mathbf{P}_n$ *is an orthogonal projection onto an approximation space* E_n*, described in* §5.4*, with components* $\mathbf{P}^1_n$ *and* $\mathbf{P}^2_n$.

Proof. We shall consider the estimation of v by v_h; the other case is similar. Similar to the proof of the previous lemma, there is a preliminary reduction to the case $\bar{v} = \bar{v}_I$ in the defining equation (5.11) for v. We consider this now. The argument is subtle. Denoting the solutions of the corresponding gradient equations, with boundary data $\bar{v}, \bar{v}_I$, respectively, by

$$v = v_\phi = \phi + \bar{v}, \quad v_\psi = \psi + \bar{v}_I, \quad \phi, \psi \in H^1_{0,\Sigma_D}(G), \tag{5.38}$$

we have

$$\begin{aligned} &\langle \exp(U - (\phi + \psi)/2 - \bar{v})\ \nabla(\phi - \psi), \nabla(\phi - \psi)\rangle = \\ &\langle \exp(U - (\phi + \psi)/2 - \bar{v})\ \nabla(\bar{v}_I - \bar{v}), \nabla(\phi - \psi)\rangle \\ -&\langle \{\exp(U - (\phi + \psi)/2 - \bar{v}) - \exp(U - (\phi + \psi)/2 - \bar{v}_I)\}\ \nabla v_\psi, \nabla(\phi - \psi)\rangle. \end{aligned} \tag{5.39}$$

Here we have used the identity,

$$\begin{aligned} \langle \exp(U - (\phi + \psi)/2 - \bar{v})\nabla v_\phi, \nabla(\phi - \psi)\rangle &= \\ 2\langle \exp(U - \phi - \bar{v})\nabla v_\phi, \nabla[\exp((\phi - \psi)/2) - 1]\rangle &= 0, \end{aligned} \tag{5.40}$$

with a similar identity for v_ψ. Note the validity of the test function choice; it vanishes on the Dirichlet boundary. The first term on the right hand side of (5.39) is easily estimated by a properly weighted sum of squares of L_2 gradient norms; one of these terms is absorbed in the usual way on the left hand side, while the other, involving interpolation error, is reflected in the estimate of the lemma. The second term is estimated by the generalized Hölder inequality; for $N = 3$, for example, as the product of L_6, L_3, and L_2 norms, respectively. The L_6 norm of the exponential difference is estimated, via Sobolev's inequality, by a constant multiple of

$$\|\bar{v} - \bar{v}_I\|_{H^1}.$$

For the critical reader, this was why the mean of ϕ and ψ was selected in the opening identity (5.39); otherwise, other terms appear which cannot be estimated. If the product of the two leading L_6 and L_3 terms, with the L_2 term, is estimated by a sum of squares, one absorbs a term on the left hand side, while retaining the term in the estimate of the lemma on the right hand side. The cases of $N = 1, 2$ are similar, and the reduction is complete. Note that the maximum principles have been used implicitly to provide pointwise bounds in the estimation. This completes the reduction to $\bar{v} = \bar{v}_I$. We shall continue to use, with each reduction, the symbol v.

The proof proceeds by a second reduction, to the case $U = U_h$, via subtraction of the weak relations defining v and $\tilde{v}_h$. The latter is the solution corresponding to the continuous problem (not the finite element formulation), with $U = U_h$. This yields

$$\langle \exp(U_h - (\tilde{v}_h + v)/2)\nabla(v - \tilde{v}_h), \nabla(v - \tilde{v}_h)\rangle =$$

$$-\langle\{\exp(U - (\tilde{v}_h + v)/2) - \exp(U_h - (\tilde{v}_h + v)/2)\}\nabla v, \nabla(v - \tilde{v}_h)\rangle,$$

by use of an identity comparable to (5.40). Hölder's inequality, as applied in the first part of the proof, followed by a sum of weighted squares, gives the term involving $U - U_h$. We chose to employ U_h, rather than U, on the left hand side of the inequality, in order to use the known regularity of ∇v.

The third and final reduction is the most demanding in terms of hypotheses. In this reduction, where we still remain in the category of solutions to continuous (not discrete) weak formulations, we solve with the replacement $\exp(-v) \mapsto \exp(-v_h)$; we denote the solution of the *linear* reduced problem by the symbol v_h^*. Thus, v_h^* solves the equation,

$$\int_G \exp(U_h - v_h)\nabla v_h^* \cdot \nabla\phi = 0, \forall\phi \in H^1_{0,\Sigma_D}(G),\ v_h^* - \bar{v}_I|_{\Sigma_D(G)} = 0. \quad (5.41)$$

Subtraction of weak relations gives

$$\int_G \exp(U_h - v_h)\nabla(v - v_h^*)\cdot \nabla(v - v_h^*) = \int_G \exp(U_h - v_h)\nabla v \cdot \nabla(v - v_h^*)$$

$$= \int_G \{\exp(U_h - v_h) - \exp(U_h - v)\}\nabla v \cdot \nabla(v - v_h^*)$$

$$\leq C\{\|v - v_h^*\|_{L_{6-\epsilon_1}} + \|v_h^* - v_h\|_{L_{6-\epsilon_1}}\}\|\nabla v\|_{L_{3+\epsilon_2}}\|\nabla(v - v_h^*)\|_{L_2}, \quad (5.42)$$

where $\epsilon_1 = \frac{4\epsilon_2}{1+\epsilon_2}$, and $\epsilon_2 > 0$ is arbitrary, subject to Assumption 5.3. The last relation is especially designed for the case $N = 3$. The reason for the employment of $\epsilon_i, i = 1, 2$, is the need to absorb the critical term involving $v - v_h^*$ from the right to the left hand side. It arises in two products, the first of which is more delicate. To do this, we may use a sharper version of the Sobolev embedding theorem. This may be written, for $N = 3$, and for functions f vanishing on Σ_D,

$$\|f\|_{L_{6-\epsilon_1}} \leq C(d)\|\nabla f\|_{H^1}. \quad (5.43)$$

Here d is the diameter of G, and $C(d) \to 0$ as $d \to 0$. Inequality (5.43), for $C(d)$ with an approximate value of 13.45 when G is a cube of side 2, is proven as a special case of [1, Lemma 4.10], in the less sharp form where the H^1 norm substitutes for gradient quantities. However, use of the equivalent H^1 norm (5.14) yields only the gradient part for functions with zero trace on Σ_D. The step from this inequality, and (5.42), to

$$C(d)C\|\nabla v\|_{L_{3+\epsilon_2}} < \exp(\gamma - \beta),$$

arises from a change of scale argument, involving shrinking G to sufficiently small diameter. This change of scale argument fails, however, if the L_6 norm is employed; this explains the introduction of ϵ_1, and the corresponding need for ϵ_2 in the Hölder inequality. One obtains, for G of sufficiently small diameter, the inequality,

$$\int_G \exp(U_h - v_h)\nabla(v - v_h^*)\cdot \nabla(v - v_h^*) \le C_1\|\nabla(v_h^* - v_h)\|_{L_2}^2. \tag{5.44}$$

Note that we have absorbed yet another term on the left hand side via a sum of squares. Since v_h can be thought of as solving a linear problem relative to v_h^*, i.e., $v_h = \mathbf{P}_n^1 v_h^*$, the statement of the lemma follows easily for domains of sufficiently small diameter. A change of the scaling variable concludes the proof for $N = 3$. The other cases are technically simpler.

5.3.2 Convergence Properties of $\mathbf{T}_n$ in Energy Norms

On the basis of (5.33, 5.37), and the properties of $\mathbf{Q}_h$, we may assume that there exists an approximation order for U_h, v_h, and w_h:

$$\|U - U_h\|_{H^1} \le Ch^\theta, \ \|v - v_h\|_{H^1} \le Ch^\theta, \ \|w - w_h\|_{H^1} \le Ch^\theta, \tag{5.45}$$

for some constant C and $0 < \theta \le 1$. Here, θ is the regularizing index introduced in Assumption 5.1 of §5.1. Note that the second term on the right-hand side of (5.33) is of order h^2, as remarked earlier.

Remark 5.3.2. In drawing this conclusion from the lemmas of the previous subsection, and general finite element approximation theory, we are assuming that no loss of assumed regularity holds for the reduced problems, employed in replacing the boundary datum by its interpolation. There are technical ways of *circumventing* this issue, such as making the common device *hypothesis* that the Dirichlet datum is constant on any given contact, i.e., on any Dirichlet boundary component.

We now provide a description of the approximation properties of $\mathbf{T}_n$. We have the following result .

Theorem 5.3.1. *The estimate,*

$$\|(\mathbf{T} - \mathbf{T}_n)[v, w]\|_{H^1 \times H^1} \le Ch^\theta, \tag{5.46}$$

holds for some constant C, uniformly over the domain Ω on which $\mathbf{T}$ *and* $\mathbf{T}_n$ *are defined. The approximation estimates* (5.45) *are assumed as described earlier.*

Proof. The proof is a routine application of the estimate (5.45), once the identity,

$$\mathbf{T}[v,w] - \mathbf{T}_n[v,w] = [\mathbf{V}(U) - \mathbf{V}_h(U_h), \mathbf{W}(U) - \mathbf{W}_h(U_h)], \tag{5.47}$$

is noted. Here we have notationally suppressed the action of the truncation map.

5.3.3 Convergence Properties of $\mathbf{T}_n$ in the Pointwise Norm

In order to obtain pointwise estimates, required for the application of the Krasnosel'skii operator calculus, we are led to the next major assumption of the chapter.

Assumption 5.3. For $N \geq 2$, the inequality,

$$\theta > N(\frac{1}{2} - \frac{1}{N}), \tag{5.48}$$

holds. Equivalently, $\theta > N(\frac{1}{2} - \frac{1}{p}),\quad p > N$. Here, θ has been defined in Assumption 5.1.

Remark 5.3.3. Suppose $f \in H^{1+\theta}$ is one of the functions u, v, w of the previous subsection, and f_h one of the corresponding functions u_h, v_h, w_h. The critical estimate, leading to the desired pointwise inequalities, is that

$$\|f_h - f_I\|_{W_p^1} \leq Ch^{\theta - N(\frac{1}{2} - \frac{1}{p})}\|f\|_{H^{1+\theta}}, \tag{5.49}$$

if $p > N$. Here, f_I is the well-defined interpolant under Assumption 5.3, and $f_h - f_I$ vanishes on the Dirichlet boundary. The estimates described earlier show that $\|f_h - f_I\|_{H^1}$ is of the order, $h^\theta\|f\|_{H^{1+\theta}}$. In order to deduce (5.49) from this estimate, we follow a suggestion of Kerkhoven; this has also been derived in the author's monograph (cf. [64, Prop. 4.2.6]). Specifically, in each simplex, the gradient of $f_h - f_I$ is constant, so that the L_p gradient norm on this element is exactly this constant, multiplied by the pth root of the simplicial volume, i.e., by a number proportional to $h^{\frac{N}{p}}$. Since the simplicial decomposition is quasi-uniform, the estimate for the L_p gradient norm can be assembled over each element, and expressed as an l_p norm of the (vector) constants on each simplex, with a factor $h^{\frac{N}{p}}$. Jensen's inequality, $\|c\|_{l_p} \leq \|c\|_{l_2}$, followed again by an L_2 gradient estimate, finally yields (5.49). Note that the gradient estimate suffices since the function $f_h - f_I$ vanishes on the Dirichlet boundary. A standard direct approximation theory estimate (cf. [64, §4.1]) yields

$$\|f - f_I\|_{W_p^1} \leq Ch^{\theta - N(\frac{1}{2} - \frac{1}{p})}\|f\|_{H^{1+\theta}}. \tag{5.50}$$

It follows that f_h converges in L_∞ to $f \in H^{1+\theta}(G)$, with order h^ρ, $\rho = \theta - N(\frac{1}{2} - \frac{1}{p})$, via the Sobolev embedding theorem.

The lemmas of §5.3.1 show the requisite convergence order in the energy norm of the composition mappings; the discussion immediately above gives the result for pointwise convergence . Similar results hold for $\mathbf{T}_n$. We have established the following.

Theorem 5.3.2. *The pointwise estimates, for $f \in H^{1+\theta}(G)$,*

$$\|f - f_h\|_{L_\infty} \leq C_1 h^\rho, \tag{5.51}$$

$$\|f - \mathbf{T}_n^i f\|_{L_\infty} \leq C_2 h^\rho, \; i = 1, 2, \tag{5.52}$$

hold for constants C_1 and C_2 and for ρ as given above. The constants depend only upon a bound for $\|f\|_{H^{1+\theta}}$. When $N = 1$, $\rho = \theta = 1$.

5.4 A Calculus for the System Mappings

We set $E = \prod_1^2 H^1(G)$ and $E_n =$ (linear) $Sp\{\bar{v}_I, S_h\}\otimes$ (linear) $Sp\{\bar{w}_I, S_h\}$, with $\mathbf{P}_n$ the orthogonal projection onto E_n. The domain Ω of the map $\mathbf{T}$ has been defined in tandem with the composition maps defining $\mathbf{T}$ in §5.1. $\mathbf{T}_n$ has been defined in §5.2, also on Ω; for consistency with Chap. 6, we shall at times restrict $\mathbf{T}_n$ to $\Omega_n = E_n \cap \Omega$. Fixed points of $\mathbf{T}$ were demonstrated to exist in [67], and were also discussed in Chap. 4; parallel arguments yield fixed points of $\mathbf{T}_n$ as well, via an application of Brouwer's fixed point theorem applied to $\bar{\Omega} \cap E_n$. The continuity of this map follows from the continuity of $\mathbf{U}_h$, $\mathbf{V}_h$, and $\mathbf{W}_h$, since $\mathbf{Tr}$ is seen to be continuous from elementary considerations; the continuity of the former mappings is studied in the next section, §5.5. We may assume, then, the existence of fixed points of $\mathbf{T}_n$.

An important approximation property of $\mathbf{P}_n$ on the union of the convex hull, *co* $R_\mathbf{T}$, of the range of $\mathbf{T}$, with $\prod_1^2 H^{1+\theta}(G) \cap H^1_{0,\Sigma_D}(G)$, is

$$\|\mathbf{P}_n\tau - \tau\|_{H^1\times H^1} \leq ch^\theta, \qquad \|\tau\|_{H^{1+\theta}\times H^{1+\theta}} \leq 1. \tag{5.53}$$

This is a consequence of the approximation theory of §5.3. The reason $\mathbf{P}_n$ does not involve affine projection is for consistency with the abstract calculus discussed in Chap. 6, where linear, not affine, subspaces are employed.

In the two subsections below we shall discuss the continuity and compactness properties of $\mathbf{U}, \mathbf{V}, \mathbf{W}$, and the properties of their Fréchet derivatives. In the subsequent section, §5.5, we shall discuss these properties for $\mathbf{U}_h$, $\mathbf{V}_h$ and $\mathbf{W}_h$. We shall have frequent cause to refer, particularly in the L_∞ estimates, as well as the linearized quasi-Fermi level problems, to the Moser regularization theory as set forth in [48, Chap. 8]. The results we shall require are for the mixed boundary-value problem, and the results we shall quote are valid for this more general situation. Also, in all cases, the underlying hypotheses (8.5–8.8) of [48] will be met, and can be checked by the reader.

5.4.1 The Map U: Differentiability Properties

We begin with a result which is a slight sharpening of Lemma 4.1 of [67]. It will be used in the proof of Lemma 5.4.2. The domain of $\mathbf{U}$ in both of these lemmas is that specified previously. The statements about Fréchet derivatives may appear purely formal in the subsections to follow. Implicit in these results, however, is the known fact that C^1 Gâteaux differentiability, with a family of bounded linear operators in the range, implies C^1 Fréchet differentiability (cf. [132]). The existence of the former property is rigorously demonstrated. The derivatives employed are actually derivatives of composition maps not restricted by pointwise constraints. Thus, the perturbations utilized in computing directional (Gâteaux) derivatives need only be in the interior of a ball defined by an energy inequality, which is equivalent to computing derivatives of natural extension maps, and, for $\mathbf{U}$, one could even extend to L_2 subsets. This is implicit in what follows, though never actually specified. We shall not distinguish notationally, but use the derivative symbol of the composition maps with their domains as previously defined. Since, in the formation of $\mathbf{T}$, the truncation map precedes the remaining composition maps, the approaches are totally equivalent.

Lemma 5.4.1. *Let* $\mathbf{U} : (v, w) \mapsto u$ *be the mapping defined implicitly through the solution of the boundary-value problem,*

$$\langle \nabla u, \nabla \phi \rangle + \langle e^{u-v} - e^{w-u} - k_1, \phi \rangle = 0, \tag{5.54}$$

where $\phi \in H^1_{0,\Sigma_D}$, *subject to the specified mixed boundary conditions in* N *dimensions. Then* $\mathbf{U}$ *is continuous from* $\prod_1^2 L_2$ *into* L_2. *In fact, it is Lipschitz continuous into* H^1. *The mapping is also Lipschitz continuous from* $\prod_1^2 L_\infty$ *to* L_∞ *if* $N \leq 5$.

Proof. The only part of the statement not covered by [67] is the L_∞ Lipschitz continuity; identity (4.1) of this reference yields the stated continuity in Hilbert space norms. As we shall cite several times in this section, the technique of proof is Moser regularization (cf. [48, Chap. 8, Theorem 8.16]), applied to the difference of two solutions of (5.54). The algebraic details are similar to the energy estimation case, and use an analog of [67, (4.1), p.578] to deduce an estimator right hand side difference in $L_{10/3}$, via the Sobolev inequality applied to the already derived continuity into H^1. Since $\frac{10}{3} > \frac{5}{2}$, as required by Moser's theory, the continuity into L_∞ follows.

Lemma 5.4.2. *Let* $\mathbf{U} : (v, w) \mapsto u$ *be the mapping defined through the solution of the boundary-value problem* (5.54). *The derivative,* $D_{(v,w)}\mathbf{U}(v, w) :$ $(\sigma, \tau) \mapsto \mu$, *is defined through the solution of the boundary-value problem,*

$$\langle \nabla \mu, \nabla \phi \rangle + \langle e^{u-v}[\mu - \sigma] + e^{w-u}[\mu - \tau], \phi \rangle = 0, \tag{5.55}$$

where $\mu|_{\Sigma_D} \equiv 0$, $\phi|_{\Sigma_D} \equiv 0$, and for all (v, w) is a uniformly bounded linear mapping from $\prod_1^2 L_2$ to H^1_{0,Σ_D}, and, in particular, compact from $\prod_1^2 L_2$ into L_q, $q < [1/2 - 1/N]^{-1}$ if $N \geq 3$ and $q < \infty$ if $N = 2$. The range of the derivative is also uniformly bounded as a family of linear mappings from $\prod_1^2 L_2$ to L_∞ when $N \leq 3$. Moreover, the mapping $(v, w) \mapsto D_{(v,w)}\mathbf{U}(v, w)$ is Lipschitz continuous from $\prod_1^2 H^1$ to the mappings from $\prod_1^2 L_2$ to H^1_{0,Σ_D} if $N \leq 4$, and is Lipschitz continuous from $\prod_1^2 H^1$ to the mappings from $\prod_1^2 L_4$ to L_∞ if $N \leq 3$.

Proof. Assume that the derivative $D_{(v,w)}\mathbf{U}$ evaluated at the point (v_1, w_1) maps the pair (σ_1, τ_1) of functions to the function μ_1, while the same derivative evaluated at the point (v_2, w_2) maps the pair (σ_2, τ_2) of functions to the function μ_2. Then

$$\langle \nabla\mu_2, \nabla(\mu_2 - \mu_1)\rangle + \langle e^{u_2 - v_2}[\mu_2 - \sigma_2] + e^{w_2 - u_2}[\mu_2 - \tau_2], (\mu_2 - \mu_1)\rangle = 0, \quad (5.56)$$

$$\langle \nabla\mu_1, \nabla(\mu_2 - \mu_1)\rangle + \langle e^{u_1 - v_1}[\mu_1 - \sigma_1] + e^{w_1 - u_1}[\mu_1 - \tau_1], (\mu_2 - \mu_1)\rangle = 0. \quad (5.57)$$

We find by subtraction:

$$0 = \|\nabla(\mu_2 - \mu_1)\|^2_{L_2} +$$
$$\langle e^{u_2 - v_2}[\mu_2 - \sigma_2] - e^{u_1 - v_1}[\mu_1 - \sigma_1] + e^{w_2 - u_2}[\mu_2 - \tau_2] - e^{w_1 - u_1}[\mu_1 - \tau_1], (\mu_2 - \mu_1)\rangle.$$

This yields, after reordering,

$$\begin{aligned}
&\|\nabla(\mu_2 - \mu_1)\|^2_{L_2} + \langle e^{u_2 - v_2}[\mu_2 - \mu_1], (\mu_2 - \mu_1)\rangle + \langle e^{w_2 - u_2}[\mu_2 - \mu_1], (\mu_2 - \mu_1)\rangle \\
&\quad = \langle [e^{u_1 - v_1} - e^{u_2 - v_2}]\mu_1, [\mu_2 - \mu_1]\rangle + \langle [e^{w_1 - u_1} - e^{w_2 - u_2}]\mu_1, [\mu_2 - \mu_1]\rangle - \\
&\qquad \langle [e^{u_1 - v_1} - e^{u_2 - v_2}]\sigma_2, [\mu_2 - \mu_1]\rangle - \langle e^{u_1 - v_1}[\sigma_1 - \sigma_2], [\mu_2 - \mu_1]\rangle - \\
&\qquad \langle [e^{w_1 - u_1} - e^{w_2 - u_2}]\tau_2, [\mu_2 - \mu_1]\rangle - \langle e^{w_1 - u_1}[\tau_1 - \tau_2], [\mu_2 - \mu_1]\rangle.
\end{aligned}$$

But each of the terms on the right hand side can be estimated through Hölder's inequality and the intermediate value theorem for the function $x \mapsto e^x$. This yields:

$$\begin{aligned}
&\|\nabla(\mu_2 - \mu_1)\|^2_{L_2} + \langle e^{u_2 - v_2}[\mu_2 - \mu_1], (\mu_2 - \mu_1)\rangle + \langle e^{w_2 - u_2}[\mu_2 - \mu_1], (\mu_2 - \mu_1)\rangle \\
&\quad \leq e^{u_{\max} - v_{\min}}\|u_1 - v_1 - (u_2 - v_2)\|_{L_p}\|\mu_1\|_{L_q}\|\mu_2 - \mu_1\|_{L_r} + \\
&\quad e^{w_{\max} - u_{\min}}\|w_1 - u_1 - (w_2 - u_2)\|_{L_p}\|\mu_1\|_{Lq}\|\mu_2 - \mu_1\|_{L_r} + \\
&\quad e^{u_{\max} - v_{\min}}\|u_1 - v_1 - (u_2 - v_2)\|_{L_p}\|\sigma_2\|_{L_q}\|\mu_2 - \mu_1\|_{L_r} + \\
&\quad e^{w_{\max} - u_{\min}}\|w_1 - u_1 - (w_2 - u_2)\|_{L_p}\|\tau_2\|_{L_q}\|\mu_2 - \mu_1\|_{L_r} + \\
&\quad e^{u_{\max} - v_{\min}}\|\sigma_1 - \sigma_2\|_{L_s}\|\mu_2 - \mu_1\|_{L_t} + e^{w_{\max} - u_{\min}}\|\tau_1 - \tau_2\|_{L_s}\|\mu_2 - \mu_1\|_{L_t},
\end{aligned} \quad (5.58)$$

where $1/p + 1/q + 1/r = 1$, and $1/s + 1/t = 1$. Now select $s = t = 2$, and $v_1 = v_2$, $w_1 = w_2$, so that $u_1 = u_2$. The uniform boundedness from $\prod_1^2 L_2$ to H^1, and hence the compactness as stated, follows for the mapping

$D_{(v,w)}\mathbf{U}(v,w) : (\sigma,\tau) \mapsto \mu$; here we take $\mu = \mu_2$, $\mu_1 = 0$, $\sigma = \sigma_2$, $\sigma_1 = 0$, $\tau = \tau_2$, $\tau_1 = 0$. The result for $N \leq 3$, of L_∞ boundedness of μ in terms of L_2 bounds on (σ, τ), follows from the Moser regularization theory, as applied to (5.55), when a general test function is employed.

To prove the results for $(v,w) \mapsto D_{(v,w)}\mathbf{U}(v,w)$, set $\sigma_1 = \sigma_2 = \sigma$, $\tau_1 = \tau_2 = \tau$, and specify $p = 4$, $q = 2$, $r = 4$ for $N \leq 4$. By the definition of the norm of a mapping, we may assume $\|\sigma\|^2_{L_2} + \|\tau\|^2_{L_2} \leq 1$, so that, by the first part of the lemma, the H^1 norm of μ_1 is uniformly bounded with respect to (v,w). If the result of Lemma 5.4.1 is combined with the Sobolev embedding theorem, $H^1 \mapsto L_4$, the lemma follows. Here we have used the fact that the left hand side of the inequality (5.58) is equivalent to the (squared) H^1 norm. The final statement is again proved by appeal to the Moser regularization theorem, as applied to the difference of (5.56) and (5.57), when a general test function is employed, in the second argument position, in place of $\mu_2 - \mu_1$. The hypotheses guarantee that the right hand side estimator is small in L_q, for $q > N/2$, as the Moser theory requires. Therefore, $q = 2$ suffices for the regularity of the estimator.

5.4.2 The Mappings V and W: Differentiability Properties

In this subsection, the domains of $\mathbf{V}$ and $\mathbf{W}$ are as defined earlier. We begin with a continuity result.

Lemma 5.4.3. *For $i = 1, 2$, and $u_i \in H^1$ satisfying the bounds* (5.10), *and the boundary conditions, let v_i be the corresponding images under the map $\mathbf{V}$, so that they appear as solutions to the weak formulation of the mixed boundary-value problem,*

$$\nabla \cdot (e^{u_i - v_i} \nabla v_i) = 0, \tag{5.59}$$

on G. Then

$$\int_G e^{u_1 - v_1} |\nabla(v_2 - v_1)|^2 dx \leq \int_G e^{u_1 - v_1} |\nabla(u_2 - u_1)|^2 dx, \tag{5.60}$$

and therefore the mapping $\mathbf{V}$ from u to v defined through (5.59) *is Lipschitz continuous, from H^1 to itself on the range of $\mathbf{U}$. The mapping is also Lipschitz continuous, from L_∞ to L_∞ on the range of $\mathbf{U}$, if Assumption* 5.3 *of* §5.3.3 *holds. Comparable statements hold for $\mathbf{W}$.*

Proof. The H^1 Lipschitz continuity has been proven by Kerkhoven. Specifically, his proof of (5.60) will appear elsewhere. The proof in the L_∞ case uses the weak relations corresponding to (5.59) above, but converted to the Slotboom variable formalism. Thus, with

$$V_i = \exp(-v_i), \quad i = 1, 2,$$

we obtain the relation,

$$\langle \exp(u_1)\nabla(V_1 - V_2), \nabla\phi\rangle = \langle(\exp(u_2) - \exp(u_1))\nabla V_2, \nabla\phi\rangle, \qquad (5.61)$$

from which it follows, via Assumption 5.3 and the Moser regularization theory, that $V_1 - V_2$ is estimated in L_∞ by a constant times an L_∞ estimate of $u_1 - u_2$. This translates to the corresponding estimate of $v_1 - v_2$. Here we have used the 'a priori' L_∞ bounds embedded in the definitions of the domains and ranges, and the L_q property of ∇V_2, for $q > N$. The proof is completed.

Remark 5.4.1. The derivative, $D\mathbf{V}(u) : \mu \mapsto \sigma$, is defined through solution of the boundary-value problem,

$$\langle e^{u-v}[(\mu - \sigma)\nabla v + \nabla\sigma], \nabla\phi\rangle = 0, \qquad (5.62)$$

where $\mu \in H^1_{0,\Sigma_D}$ and $\sigma \in H^1_{0,\Sigma_D}$. Here, ϕ is a test function in H^1_{0,Σ_D}. For smooth v, the standard existence theory (cf. [48, Chap. 8, Th. 8.3]) yields a solution σ for $N \le 3$. The lemma to follow, in which H^1 bounds for σ are determined in terms of μ, allows limits of v and σ, and hence solutions for v with regularity specified by [106]. Finally, under the more enhanced regularity of Assumption 5.3, the Moser regularization theory (cf. [48, Chap. 8]) gives 'a priori' L_∞ bounds on σ in terms of $H^{1+\theta}$ bounds (less stringently, W^1_p bounds, $p > N$) on v and L_∞ bounds on μ. This is required in the proof of Lemma 5.4.5. Moreover, in this case, it is then immediate from the relation (5.62) that $\nabla\sigma \in L_q, q > N$, with a bound in this norm given by an L_∞ bound for μ, and an $H^{1+\theta}$ bound for v.

Lemma 5.4.4. *Let $u \in H^1 \cap L_\infty$ be given and, for $N \le 3$, let v be the solution to the weak formulation of the mixed boundary-value problem,*

$$\nabla \cdot (e^{u-v}\nabla v) = 0,$$

on G. Then the derivative $D\mathbf{V}$ of the mapping $\mathbf{V}$ from u to v defined through this equation is uniformly bounded from H^1_{0,Σ_D} to itself, satisfying the inequality,

$$e^{(u_{\max} - v_{\min})/2}\|\nabla\mu\|_{L_2} \ge e^{(u_{\min} - v_{\max})/2}\|\nabla\sigma\|_{L_2},$$

and hence, for each u, is compact from H^1_{0,Σ_D} into L_2. Here, μ and σ are defined in Remark 5.4.1 *above.*

Proof. The proof is due to Kerkhoven, and will appear elsewhere.

Lemma 5.4.5. *Suppose Assumption* 5.3 *holds. Then the derivative $D\mathbf{V}$, as defined by equation* (5.62), *is a locally Lipschitz continuous mapping from H^1 to the mappings from $H^1 \cap L_\infty$ to H^1_{0,Σ_D} for $N \le 3$. A similar statement holds for $\mathbf{W}$. The derivative is also a locally Lipschitz continuous mapping from L_∞ to the mappings from $H^1 \cap L_\infty$ to L_∞.*

Proof. We estimate

$$\sigma_2 - \sigma_1 \equiv [D\mathbf{V}(u_2) - D\mathbf{V}(u_1)]\mu,$$

as defined through the equations,

$$\langle e^{u_2-v_2}[(\mu-\sigma_2)\nabla v_2 + \nabla\sigma_2], \nabla\phi\rangle = 0, \tag{5.63}$$

and

$$\langle e^{u_1-v_1}[(\mu-\sigma_1)\nabla v_1 + \nabla\sigma_1], \nabla\phi\rangle = 0. \tag{5.64}$$

We start by writing

$$\langle e^{u_2-v_2}\nabla(\sigma_2-\sigma_1), \nabla(\sigma_2-\sigma_1)\rangle = \langle e^{u_2-v_2}\nabla\sigma_2, \nabla(\sigma_2-\sigma_1)\rangle -$$
$$\langle e^{u_1-v_1}\nabla\sigma_1, \nabla(\sigma_2-\sigma_1)\rangle + \langle [e^{u_1-v_1} - e^{u_2-v_2}]\nabla\sigma_1, \nabla(\sigma_2-\sigma_1)\rangle.$$

Now we employ (5.63, 5.64) to rewrite the first and second terms, so that we obtain

$$\langle e^{u_2-v_2}\nabla(\sigma_2-\sigma_1), \nabla(\sigma_2-\sigma_1)\rangle = \langle e^{u_2-v_2}(\sigma_2-\mu)\nabla v_2, \nabla(\sigma_2-\sigma_1)\rangle -$$
$$\langle e^{u_1-v_1}(\sigma_1-\mu)\nabla v_1, \nabla(\sigma_2-\sigma_1)\rangle + \langle [e^{u_1-v_1} - e^{u_2-v_2}]\nabla\sigma_1, \nabla(\sigma_2-\sigma_1)\rangle. \tag{5.65}$$

But on the right hand side we can write

$$\langle e^{u_2-v_2}(\sigma_2-\mu)\nabla v_2, \nabla(\sigma_2-\sigma_1)\rangle =$$
$$\langle e^{u_2-v_2}(\sigma_2-\sigma_1)\nabla v_2, \nabla(\sigma_2-\sigma_1)\rangle + \langle e^{u_2-v_2}(\sigma_1-\mu)\nabla v_2, \nabla(\sigma_2-\sigma_1)\rangle =$$
$$\int_G e^{u_2-v_2}\nabla v_2 \cdot \nabla\frac{1}{2}(\sigma_2-\sigma_1)^2 dx + \langle e^{u_2-v_2}(\sigma_1-\mu)\nabla v_1, \nabla(\sigma_2-\sigma_1)\rangle +$$
$$\langle e^{u_2-v_2}(\sigma_1-\mu)\nabla(v_2-v_1), \nabla(\sigma_2-\sigma_1)\rangle.$$

The integral in the preceding equation is equal to 0 by the definition of $\mathbf{V}$, and therefore (5.65) yields

$$\langle e^{u_2-v_2}\nabla(\sigma_2-\sigma_1), \nabla(\sigma_2-\sigma_1)\rangle = \langle [e^{u_1-v_1} - e^{u_2-v_2}]\nabla\sigma_1, \nabla(\sigma_2-\sigma_1)\rangle +$$
$$\langle [e^{u_1-v_1} - e^{u_2-v_2}](\mu-\sigma_1)\nabla v_1, \nabla(\sigma_2-\sigma_1)\rangle -$$
$$\langle e^{u_2-v_2}(\mu-\sigma_1)\nabla(v_1-v_2), \nabla(\sigma_1-\sigma_2)\rangle.$$

Each term on the right hand side may be estimated by Hölder's inequality in combination with Sobolev's inequality. For example, by Hölder's inequality,

$$|\langle [e^{u_1-v_1} - e^{u_2-v_2}]\nabla\sigma_1, \nabla(\sigma_2-\sigma_1)\rangle|$$
$$\leq \|e^{u_1-v_1} - e^{u_2-v_2}\|_{L_r}\|\nabla\sigma_1\|_{L_s}\|\nabla(\sigma_2-\sigma_1)\|_{L_2},$$

where $\frac{1}{r} + \frac{1}{s} = \frac{1}{2}$. Next, Sobolev's inequality yields

$$\leq Ce^{u_{\max}-v_{\min}}\|(u_1-u_2) + (v_1-v_2)\|_{H^1}\|\nabla\sigma_1\|_{L_s}\|\nabla(\sigma_2-\sigma_1)\|_{L_2},$$

where, for $N = 3$, Sobolev's inequality requires that $r = 6$. Therefore, we must take $s = 3$. For $N = 2$, by Sobolev's inequality, $r < \infty$ is arbitrary, and therefore $s > 2$. In either case, a bound for $\nabla\sigma_1$ is rendered via an L_∞ bound for μ and an $H^{1+\theta}$ bound for v_1, as discussed in Remark 5.4.1 following Lemma 5.4.3. Note also that the bounds of Lemma 5.4.4 give L_2 bounds on $\nabla\sigma_1$, in terms of L_2 bounds on $\nabla\mu$. By use of the inequality, $ab \leq \frac{1}{2}[\epsilon a^2 + \epsilon^{-1}b^2]$, $\epsilon > 0$, we can absorb the term $\|\nabla(\sigma_2 - \sigma_1)\|_{L_2}^2$ on the left hand side by appropriate choice of ϵ. By making use of the L_∞ bounds for $\mu - \sigma_1$, it is possible to estimate the other two terms in a similar manner, including the absorption of the term $\|\nabla(\sigma_1 - \sigma_1)\|_{L_2}^2$. Altogether, we obtain an estimate of the form,

$$\langle e^{u_2 - v_2}\nabla(\sigma_2 - \sigma_1), \nabla(\sigma_2 - \sigma_1)\rangle \leq C[\|u_1 - u_2\|_{H^1}^2 + \|v_1 - v_2\|_{H^1}^2],$$

where C depends on $\|v_1\|_{H^1}$, on the 'a priori' L_∞ bounds for u_i, v_i, and on an 'a priori' $L_\infty \cap H^1$ bound for μ. The first part of the lemma follows since the argument for $\mathbf{W}$ is similar.

The L_∞ result depends upon the following identity, which was essentially derived in the first part of the proof:

$$\langle e^{u_2 - v_2}\nabla(\sigma_2 - \sigma_1), \nabla\phi\rangle - \langle e^{u_2 - v_2}(\sigma_2 - \sigma_1)\nabla v_2, \nabla\phi\rangle =$$

$$\langle [e^{u_1 - v_1} - e^{u_2 - v_2}]\nabla\sigma_1, \nabla\phi\rangle +$$

$$\langle [e^{u_1 - v_1} - e^{u_2 - v_2}](\mu - \sigma_1)\nabla v_1, \nabla\phi\rangle + \langle e^{u_2 - v_2}(\mu - \sigma_1)\nabla(v_1 - v_2), \nabla\phi\rangle.$$

According to the Moser regularization theory (cf. [48, Chap. 8, Theorem 8.16]), an L_∞ estimate for $\sigma_2 - \sigma_1$ is determined in terms of L_q estimates, $q > N$, of each of the three first argument terms on the right hand side of the identity. In the case of the first two terms, given the gradient inclusion of v_1 and σ_1 in $L_q, q > N$, it follows that an L_∞ estimate of $\sigma_2 - \sigma_1$ depends upon corresponding estimates for $u_1 - u_2$; here we may use the result of Lemma 5.4.3. The third term involves the L_q norm of $\nabla(v_1 - v_2)$, which is estimated as follows. First, a reduction to the case of estimation of $\nabla(V_1 - V_2)$, where upper case denotes Slotboom variables, is achieved by the result of Lemma 5.4.3 and Assumption 5.3. The gradient estimation of the Slotboom variables makes use of (5.61), which shows that an estimate of $u_1 - u_2$ in L_∞ suffices. This completes the proof.

5.5 The Mappings $\mathbf{U}_h$, $\mathbf{V}_h$, and $\mathbf{W}_h$

In this section we develop a differential calculus for the numerical fixed point map by examining the discrete composition mappings. This includes pertinent convergence results. These results depend upon regularization properties of $D\mathbf{U}$, $D\mathbf{V}$, and $D\mathbf{W}$. They are stated in the form of the next assumption.

Assumption 5.4 (linear regularization). We assume θ satisfies Assumption 5.3; moreover, that the mappings described by (5.55) and (5.62) are regularizing at arbitrary domain points $[v, w]$ and u. Specifically, in the first equation, a bound in $H^{1+\theta}$ for μ depends only upon L_2 bounds for σ and τ. In the second equation, a bound in $H^{1+\theta}$ for σ depends only upon an L_∞ bound for μ and an $H^{1+\theta}$ bound (less stringently, a W_p^1 bound, $p > N$) for v. A similar assumption holds for the derivative map of $\mathbf{W}$.

5.5.1 The Mapping $\mathbf{U}_h$

We begin with a continuity result for $\mathbf{U}_h$.

Lemma 5.5.1. *The mapping $\mathbf{U}_h$, defined in* §5.2.1, *is Lipschitz continuous on its domain from $\prod_1^2 L_2$ to H^1. For $N \leq 5$, it is also Lipschitz continuous from $\prod_1^2 L_\infty$ to L_∞ under Assumption* 5.3.

Proof. The proof follows that of Lemma 5.4.1 for the first statement. The second statement may be derived by applying the triangle inequality to the corresponding Lipschitz continuity of $\mathbf{U}$, derived in Lemma 5.4.1, and the pointwise convergence properties of $\mathbf{U}_h$, inferred from §5.3.3 under Assumption 5.3.

We turn now to the differentiability properties of $\mathbf{U}_h$.

Lemma 5.5.2. *The derivative $D_{(v,w)}\mathbf{U}_h(v, w) : (\sigma, \tau) \mapsto \mu_h$ is defined via the solution of the projection relation,*

$$\langle \nabla \mu_h, \nabla \phi \rangle + \langle e^{U_h - v}[\mu_h - \sigma] + e^{w - U_h}[\mu_h - \tau], \phi \rangle = 0, \tag{5.66}$$

where μ_h and ϕ are in S_h and σ and τ are in H^1_{0,Σ_D}. The mapping $(v, w) \mapsto D_{(v,w)}\mathbf{U}_h(v, w)$ is, for all (v, w), a uniformly bounded linear mapping from the topology of $\prod_1^2 L_2$ to that of H^1_{0,Σ_D}. Under Assumption 5.3, *the range of the derivative is also uniformly bounded as a family of linear mappings from the topology of $\prod_1^2 L_2$ to L_∞. Moreover, the mapping $(v, w) \mapsto D_{(v,w)}\mathbf{U}_h(v, w)$ is Lipschitz continuous from $\prod_1^2 H^1$ to the topology of the mappings from $\prod_1^2 L_2$ to H^1_{0,Σ_D} if $N \leq 4$, and, under Assumption* 5.4, *is Lipschitz continuous from $\prod_1^2 H^1$ to the topology of the mappings from $\prod_1^2 L_4$ to L_∞ if $N \leq 3$. The Lipschitz constants are independent of h.*

Proof. The proof follows that of Lemma 5.4.2, for the Hilbert space properties, and includes arguments from earlier sections for the pointwise properties.

Lemma 5.5.3. *Under Assumption* 5.4, *the solutions*

$$D_{(v,w)}\mathbf{U}(v, w)(\sigma, \tau) := \mu$$

of (5.55) *and*

$$D_{(v,w)}\mathbf{U}_h(v,w)(\sigma,\tau) := \mu_h$$

of (5.66) *satisfy an estimate of the form,*

$$\|\mu - \mu_h\|_{H^1} \leq Ch^\theta \|[\sigma,\tau]\|_{H^1 \times H^1}, \tag{5.67}$$

where C does not depend upon h, v, or w. In particular, the L_∞ estimate,

$$\|\mu - \mu_h\|_{L_\infty} \leq Ch^\rho \|[\sigma,\tau]\|_{L_\infty \times L_\infty}, \tag{5.68}$$

holds. Here ρ is introduced in §5.3.3.

Proof. It is advantageous to introduce an intermediate problem: the relation (5.55) with $u = U$ replaced by $u = U_h$, leading to a solution μ_*. The estimation of the approximation of μ by μ_* can be handled by the proof of Lemma 5.4.2, as is now described. Upon subtraction of the relations (5.55), corresponding to μ and μ_*, we can make use of the estimate (5.58), with the formal identifications, $\mu_1 = \mu, \mu_2 = \mu_*, \sigma_1 = \sigma_2 = \sigma, \tau_1 = \tau_2 = \tau, v_1 = v_2 = v, w_1 = w_2 = w, u_1 = U, u_2 = U_h$. The same arguments can now be used as in the proof of Lemma 5.4.2 to deduce the inequality,

$$\|\mu - \mu_*\|_{H^1} \leq C_1 \|U - U_h\|_{H^1} \quad \|[\sigma,\tau]\|_{H^1 \times H^1},$$

from which the correct estimate in (5.67) follows for this term in the triangle inequality. The remaining step is the estimation of the approximation of μ_* by μ_h, via the characterization of μ_h as a standard finite element approximation. Here it is necessary to utilize Assumption 5.4, yielding enhanced regularity of μ_*, and then the usual finite element estimation, as developed in §5.3, for the linear problem with mixed boundary conditions. As usual, the maximum principles, embedded in the domain definitions, ensure that the exponential multiplier is uniformly bounded away from zero. The finite element estimate is consistent with (5.67), and even employs a weaker (L_2) norm for σ and τ. This completes the verification of (5.67). The relation (5.68) follows as in §5.3.3. This completes the proof.

5.5.2 The Mappings $\mathbf{V}_h$ and $\mathbf{W}_h$

In this subsection, we shall establish convergence properties associated with the differentiated quasi-Fermi level mappings, necessary for an application of the general calculus. It will be advantageous to rework the framework of the equations defining $D\mathbf{V}$ and $D\mathbf{W}$, in order to study the corresponding projections adequately. However, we shall first state our final assumption, required for a convergence analysis of the differentiated maps. This assumption is consistent with the stability results discussed in [17, Chapter 7].

Assumption 5.5 (finite element linear stability). For θ given by Assumption 5.3, the finite element approximations v_h and w_h are well defined and stable in $W_p^1, p > N$. Moreover, the quadratic form $A_h(\cdot,\cdot)$, defined from (5.70) below, has a positive minimum on the finite element space, S_h.

Consistent with the previous pattern, we shall specifically analyze only the map $D\mathbf{V}$ and its discretization, and the statements of the results will be reserved for this case for purposes of economy.

Definition 5.5.1. *Given u, v, v_h, we define the following bilinear forms:*

$$A(\sigma,\phi) = \langle e^{u-v}[\nabla\sigma - \sigma\nabla v], \nabla\phi\rangle, \tag{5.69}$$

$$A_h(\sigma,\phi) = \langle e^{u-v_h}[\nabla\sigma - \sigma\nabla v_h], \nabla\phi\rangle, \tag{5.70}$$

where $\sigma, \phi \in H^1_{0,\Sigma_D}$. In terms of the functional A, the weak form of the derivative relation, $D\mathbf{V}(u) : \mu \mapsto \sigma$, may be written

$$A(\sigma,\phi) = \ell(\phi), \tag{5.71}$$

where the continuous linear functional ℓ is given by

$$\ell(\phi) = -(\exp(u-v)\mu\nabla v, \nabla\phi)_{L_2},$$

and $v = \mathbf{V}(u)$. Similar remarks pertain to the case of A_h.

Remark 5.5.1. We have already discussed the existence question for σ in the remark following Lemma 5.4.3. However, additional properties are required for a finite element convergence theory. We shall base the latter upon the inf-sup theory referenced in Chap. 6 (cf. (5.83) below). We shall provide a preliminary framework of existence/uniqueness for (5.71), which will guarantee that the direct and adjoint problems are well posed, via the Fredholm alternative for compact operators. Similar well posedness applies to A_h, and is a necessary precondition for the inf-sup theory. An important relation, which lies at the center of the entire process, connects A with a well behaved bilinear form B. The relation, when well-defined, is given by

$$A(\sigma, \exp(-v)\phi) = B(\sigma,\phi), \tag{5.72}$$

where B is defined via

$$B(\sigma,\phi) = \int_G \{J_\sigma \cdot J_\phi\}\exp(-u), \tag{5.73}$$

and

$$J_\phi = \exp(u-v)(\nabla\phi - \phi\nabla v). \tag{5.74}$$

The following result is the critical one.

Lemma 5.5.4. *The norm defined by $B(\cdot\,,\cdot)^{1/2}$ is equivalent to the norm on H^1_{0,Σ_D}. Moreover, a unique solution of the operator equation,*

$$-\nabla\cdot J_\sigma = \ell, \tag{5.75}$$

exists, which may be identified with the unique solution of (5.62). *Equivalently, σ satisfies*

$$B(\sigma,\psi) = \ell(\exp(-v)\psi),\ \psi \in H^1_{0,\Sigma_D}. \tag{5.76}$$

Proof. If (5.73) is defined for functions in C^∞, which vanish on Σ_D with associated flux J vanishing on Σ_N, with this class designated in the proof by $\mathcal{C}$ and assumed dense in H^1_{0,Σ_D}, then the relation,

$$B(\phi, \psi) = (L\phi, \psi)_{L_{2,\omega}} = (\phi, L\psi)_{L_2,\omega}, \tag{5.77}$$

holds by direct computation. Here, L is defined by

$$L\phi = -\nabla \cdot J_\phi, \tag{5.78}$$

and $L_{2,\omega}$ is the weighted L_2 space, with weight, $\exp(-v)$. The relation (5.77) shows that the symmetric operator L is semibounded (nonnegative definite), hence has a self-adjoint Friedrichs' extension, $\hat{L}$ (cf. [112, pp. 329–335]). The domain $\hat{\mathcal{C}}$ of $\hat{L}$ is densely contained in a canonical extension, described below, of the smooth domain $\mathcal{C}$ of L; the tensor product of $\mathcal{C}$ with itself is the original domain of the bilinear form, $B(\cdot\,,\cdot)$. Specifically, the domain $\mathcal{C}$ of the quadratic form associated with $B(\cdot\,,\cdot)$ is completed by the usual equivalence classes of Cauchy sequences; noteworthy is the fact that the construction requires the sequences to be Cauchy with respect to $B(\cdot\,,\cdot)^{1/2}$, as well as with respect to the L_2 metric. This permits the identification of the completion set with H^1_{0,Σ_D}, and may be achieved by the standard completion with respect to $B(\cdot,\cdot) + c(\cdot,\cdot)_{L_2}$, for any positive c. A computation shows that, for c sufficiently large, the standard H^1 norm is equivalent to the norm induced by this quadratic form. In particular, the evaluation of $\hat{L}$ on its extended domain $\hat{\mathcal{C}}$ may be achieved as in (5.78). This operator is coercive with respect to L_2; this follows from a study of the compact inverse, if 0 is not an eigenvalue of $\hat{L}$. In order to demonstrate that 0 is not an eigenvalue, suppose

$$\hat{L}\zeta = 0.$$

One concludes that $\zeta = 0$, via the maximum principle result developed in [48, §8.1]. Although quoted for the full Dirichlet problem, the result is valid for the mixed problem, with (homogeneous) Dirichlet data imposed on Σ_D. The equivalence of the standard H^1 norm with that defined by $B(\cdot\,,\cdot)^{1/2}$ now follows, since the latter norm can be compared with the norm involving the perturbation by the constant c above. A simple minimization procedure, involving completion of the square and the Riesz representation theorem, leads to the existence of a unique σ satisfying (5.5.4). As the representer of $\ell \circ \exp(-v)$, σ becomes the solution in this approach (cf.[102]), and may then be identified with the solution of the derivative relation (5.62). This completes the proof.

Remark 5.5.2. The finite element approximation, associated with the derivative maps, will be defined in the following lemma as a standard Galerkin approximation. Nonstandard Petrov-Galerkin formulations could have been employed in the original definition of the nonlinear approximations v_h and w_h, but we chose to employ standard theory. This explains the introduction of the bilinear form, B, and the development of its properties, as a way to facilitate the application of the inf-sup theory. The latter is introduced, and generalized, in the following chapter.

Lemma 5.5.5. *The derivative* $\mathbf{DV}_h(u) : \mu \mapsto \sigma_h$ *may be defined by*

$$\langle e^{u-v_h}[(\mu - \sigma_h)\nabla v_h + \nabla \sigma_h], \nabla \phi \rangle = 0, \tag{5.79}$$

where $\sigma_h, \phi \in S_h$. *Under Assumptions* 5.4 *and* 5.5, *the solutions* $\mathbf{DV}(u)(\mu) := \sigma$ *of* (5.62) *and* $\mathbf{DV}_h(u)(\mu) := \sigma_h$ *of* (5.79) *satisfy an estimate of the form,*

$$\|\sigma - \sigma_h\|_{H^1} \leq Ch^\theta \|\mu\|_{L_\infty \cap H^1}, \tag{5.80}$$

where C *does not depend upon* h *or* u*; note that* u *and* μ *are the same for each of the two maps. In particular, the* L_∞ *estimate,*

$$\|\sigma - \sigma_h\|_{L_\infty} \leq Ch^\rho \|\mu\|_{L_\infty}, \tag{5.81}$$

holds as well.

Proof. The first part of the proof uses an adaptation of Lemma 5.4.5, in order to make an intermediate reduction to the case where σ, the range element in the definition of $\mathbf{DV}(u)(\mu)$, is replaced by f_h; here, the notation f_h is understood to mean the solution of (5.62) with $v \mapsto v_h$. The existence in this case makes use of A_h. Thus, make the identifications, in the proof of the lemma,

$$\mathbf{V}(u_2) \mapsto v_h, \sigma_2 \mapsto f_h.$$

The proof of Lemma 5.4.5 and Assumption 5.3 now yield the estimate,

$$\|\sigma - f_h\|_{H^1} \leq Ch^\theta \|\mu\|_{L_\infty \cap H^1}. \tag{5.82}$$

The estimate,

$$\|f_h - \sigma_h\|_{H^1} \leq Ch^\theta \|f_h\|_{H^{1+\theta}}, \tag{5.83}$$

follows from the inf-sup saddle point theory (cf. the final part of Assumption 5.5), as outlined in §6.1, in conjunction with Assumption 5.4. Now, the statement of Assumption 5.4 specifies the dependence of $\|f_h\|_{H^{1+\theta}}$, viz., on $\|v_h\|_{W_p^1}$ and on $\|\mu\|_{L_\infty}$. The estimation of $\|v_h\|_{W_p^1}$ is effected by Assumption 5.5. This finally yields (5.80). The relation (5.81) follows as in §5.3.3. This completes the proof.

5.6 Summary of Results for T and $\mathbf{T}_n$

In this section, we shall use the results developed earlier in the chapter to obtain the necessary properties of $\mathbf{T}$ and $\mathbf{T}_n$. All of the assumptions, and hence all of the previously derived results, will be utilized, at least implicitly. We begin by describing a result for the truncation map. The routine proof is omitted.

Lemma 5.6.1. *The truncation map* $\mathbf{Tr}$, *defined by* (5.7), *is Lipschitz continuous from* $\Omega \cap \prod_1^2 L_\infty$ *to its range. Its derivative, defined by,*

$$\mathbf{Tr}'[v,w](\phi,\psi) = [\mathbf{h}_1'(v)\phi, \mathbf{h}_2'(w)\psi], \tag{5.84}$$

satisfies a similar Lipschitz property. Specifically, $\mathbf{Tr}'$ *is Lipschitz continuous from* $\Omega \cap \prod_1^2 L_\infty$ *to the mappings from* $\prod_1^2 H^1$ *to* $\prod_1^2 H^1_{0,\Sigma_D} \cap \prod_1^2 L_\infty$.

The representation of $\mathbf{T}'$ is directly given by the chain rule:

$$\begin{aligned}\mathbf{T}'(v,w) = [&\mathbf{V}'(\mathbf{U}(\mathbf{Tr}(v,w))) \circ \mathbf{U}'(\mathbf{Tr}(v,w)) \circ \mathbf{Tr}'(v,w),\\ &\mathbf{W}'(\mathbf{U}(\mathbf{Tr}(v,w))) \circ \mathbf{U}'(\mathbf{Tr}(v,w)) \circ \mathbf{Tr}'(v,w)].\end{aligned} \tag{5.85}$$

Of course, the action of the mapping just defined is expressed by application to a test function pair, in which case the derivative of the truncation map, acting chronologically first in each component, acts via multiplier application to the test functions. The fact that the derivative of $\mathbf{T}$ is to be interpreted in the sense of the Fréchet derivative follows from the chain rule; the truncation map is differentiable by direct verification, and the the remaining composition maps have been analyzed previously. The following results are to be used in the proofs of Lemmas 5.7.1 and 5.7.2, and are noted, with a summary of the verification. $N \leq 3$ is required for the statements.

1 $\mathbf{T}'$ is Lipschitz continuous from $\Omega \cap \prod_1^2 L_\infty$ to the mappings from $\prod_1^2 H^1_{0,\Sigma_D}$ to $\prod_1^2 H^1_{0,\Sigma_D} \cap \prod_1^2 L_\infty$.

This follows from the application of Lemmas 5.4.2, 5.4.4, 5.4.5, and 5.6.1 to the following identity for first components, and a similar one for second components:

$$\begin{aligned}&\mathbf{V}'(\mathbf{U}(\mathbf{Tr}(v_1,w_1))) \circ \mathbf{U}'(\mathbf{Tr}(v_1,w_1)) \circ \mathbf{Tr}'(v_1,w_1) -\\ &\mathbf{V}'(\mathbf{U}(\mathbf{Tr}(v_2,w_2))) \circ \mathbf{U}'(\mathbf{Tr}(v_2,w_2)) \circ \mathbf{Tr}'(v_2,w_2) =\\ &\mathbf{V}'(\mathbf{U}(\mathbf{Tr}(v_1,w_1))) \circ \mathbf{U}'(\mathbf{Tr}(v_1,w_1)) \circ \{\mathbf{Tr}'(v_1,w_1) - \mathbf{Tr}'(v_2,w_2)\}\\ &+\mathbf{V}'(\mathbf{U}(\mathbf{Tr}(v_1,w_1))) \circ \{\mathbf{U}'(\mathbf{Tr}(v_1,w_1)) - \mathbf{U}'(\mathbf{Tr}(v_2,w_2))\} \circ \mathbf{Tr}'(v_2,w_2)\\ &+ \{\mathbf{V}'(\mathbf{U}(\mathbf{Tr}(v_1,w_1))) - \mathbf{V}'(\mathbf{U}(\mathbf{Tr}(v_2,w_2)))\} \circ \mathbf{U}'(\mathbf{Tr}(v_2,w_2)) \circ \mathbf{Tr}'(v_2,w_2).\end{aligned} \tag{5.86}$$

2 $\mathbf{T}'$ is uniformly bounded over Ω as a family of linear mappings from $\prod_1^2 H^1_{0,\Sigma_D}$ to $\prod_1^2 H^{1+\theta} \cap \prod_1^2 H^1_{0,\Sigma_D} \cap \prod_1^2 L_\infty$.

This follows directly from Assumption 5.4, the remark following Lemma 5.4.3, and Lemmas 5.4.2 and 5.4.4.

Prior to stating the next property, we note the representation for $\mathbf{T}'_n$:

$$\begin{aligned}\mathbf{T}'_n(v,w) = [&\mathbf{V}'_h(\mathbf{U}_h(\mathbf{Tr}(v,w))) \circ \mathbf{U}'_h(\mathbf{Tr}(v,w)) \circ \mathbf{Tr}'(v,w),\\ &\mathbf{W}'_h(\mathbf{U}_h(\mathbf{Tr}(v,w))) \circ \mathbf{U}'_h(\mathbf{Tr}(v,w)) \circ \mathbf{Tr}'(v,w)]. \end{aligned} \tag{5.87}$$

A considerable simplification has resulted from the fact that each linear map in the range of the derivative is defined on the entire space, $\prod_1^2 H^1_{0,\Sigma_D}$, and not simply a finite dimensional subspace. We now present the next property.

3 $\mathbf{T}'_n$ converges uniformly to $\mathbf{T}'$. Specifically, in the uniform operator topology determined for mappings from $\prod_1^2 H^1_{0,\Sigma_D}$ into itself, $\|\mathbf{T}'(z) - \mathbf{T}'_n(z)\|$ tends to zero as $n \to \infty$, independent of $z \in \Omega$. The same statement holds in the uniform operator topology determined for mappings from $\prod_1^2 H^1_{0,\Sigma_D} \cap \prod_1^2 L_\infty$ into itself.

The result follows by application of Lemmas 5.4.5, 5.5.1–5.5.3, 5.5.5, 5.6.1, and inequality (5.45) to the following identity for first components, and a similar one for second components:

$$\begin{aligned}&\mathbf{V}'(\mathbf{U}(\mathbf{Tr}(v,w))) \circ \mathbf{U}'(\mathbf{Tr}(v,w)) \circ \mathbf{Tr}'(v,w) -\\ &\mathbf{V}'_h(\mathbf{U}_h(\mathbf{Tr}(v,w))) \circ \mathbf{U}'_h(\mathbf{Tr}(v,w)) \circ \mathbf{Tr}'(v,w) =\\ &\mathbf{V}'(\mathbf{U}(\mathbf{Tr}(v,w))) \circ \{\mathbf{U}'(\mathbf{Tr}(v,w)) - \mathbf{U}'_h(\mathbf{Tr}(v,w))\} \circ \mathbf{Tr}'(v,w)\\ &+\{\mathbf{V}'(\mathbf{U}(\mathbf{Tr}(v,w))) - \mathbf{V}'(\mathbf{U}_h(\mathbf{Tr}(v,w)))\} \circ \mathbf{U}'_h(\mathbf{Tr}(v,w)) \circ \mathbf{Tr}'(v,w)\\ &+\{\mathbf{V}'(\mathbf{U}_h(\mathbf{Tr}(v,w))) - \mathbf{V}'_h(\mathbf{U}_h(\mathbf{Tr}(v,w)))\} \circ \mathbf{U}'_h(\mathbf{Tr}(v,w)) \circ \mathbf{Tr}'(v,w).\end{aligned} \tag{5.88}$$

4 $\mathbf{T}'_n$ is continuous, uniformly in n and elements of its domain, in both uniform operator topologies described in the previous item.

This follows directly from the two previous items and the triangle inequality.

5 $\mathbf{T}'(z)$ is compact, for each $z \in \Omega$, as a mapping from $\prod_1^2 H^1_{0,\Sigma_D}$ into itself.

This follows from the boundedness assertions of Lemma 5.4.2, the regularization hypothesis inherent in Assumption 5.4, and the compact injection of $H^{1+\theta}$ into H^1.

Remark 5.6.1. The Lipschitz continuity of the derivative in Lemma 5.6.1 depends upon the joint H^1-L_∞ norm employed, and is not valid in the H^1 norm alone.

5.7 Verification of the General Hypotheses

5.7.1 Verification of the 'A Priori' Estimates

The following lemma affords a verification of the hypotheses of Theorem 6.2.1 for the semiconductor application. It is assumed throughout that the Euclidean dimension N satisfies $N \leq 3$.

Lemma 5.7.1. *The operators* $\mathbf{T}$, $\mathbf{P}_n\mathbf{T}$, *and* $\mathbf{T}_n$ *are Fréchet differentiable in* Ω, *and, in particular,* $\mathbf{T}_n$ *is Fréchet-differentiable in* Ω_n. *Define the closed subspace,* $E_* = \prod_1^2 H^1 \cap L_\infty$ *of* $\prod_1^2 H^1(G)$, *with the accompanying norm,*

$$\|f\|_* = \max\{\|f\|_{H^1 \times H^1}, \|f\|_{L_\infty \times L_\infty}\}. \tag{5.89}$$

Then the hypotheses of Theorem 6.2.1 *hold. Specifically, if* $x_0 \in \Omega$ *is a fixed point of* $\mathbf{T}$, *guaranteed by the theory of Chap. 4, then:*

1. $\mathbf{T}(\mathbf{P}_n x_0)$ *is approximated by* $\mathbf{T}_n(\mathbf{P}_n x_0)$, *in the norm* $\| \ \|_*$ *as* $n \to \infty$; $\mathbf{T}$ *is continuous in this norm; and,*
$$\|\mathbf{P}_n(x_0) - x_0\|_* \to 0.$$
2. *Moreover, there exist a positive number* δ_* *and a positive integer* n_* *such that*
$$\mathbf{P}_n x_0 \in \mathcal{U}_n := \{x \in \Omega_n : \|x - x_0\|_* \leq \delta_*\}, \quad n \geq n_*,$$
and $\mathbf{T}_n'(\mathbf{P}_n x_0)$ *is approximated by* $\mathbf{T}_n'(x)$ *on* $\mathcal{U}_n \subset E_*$, *uniformly in* x, *in the uniform operator topologies as derived from both the energy norm and the* $*$ *norm.*
3. $\mathbf{T}'$ *is continuous in* E_* *at* x_0 *with respect to the uniform operator topologies as derived from both norms.*
4. $\mathbf{P}_n\mathbf{T}'(\mathbf{P}_n x_0)$ *is approximated by* $\mathbf{T}_n'(\mathbf{P}_n x_0)$, *in the uniform operator topologies on* E_n, *as derived from both norms.*

Proof. The statements follow directly from the results of §5.4–5.6, especially the latter.

5.7.2 Verification of the 'A Posteriori' Estimates

The following lemma addresses the hypotheses of Theorem 6.2.2 for the semiconductor application. It is assumed throughout that the Euclidean dimension N satisfies $N \leq 3$. For convenience, in this subsection, we shall denote by unsubscripted norms the product H^1 norm, or associated operator norm.

Lemma 5.7.2. *Let the operators* $\mathbf{T}, \mathbf{P}_n\mathbf{T}$, *and* $\mathbf{T}_n$ *be as defined previously; the norm* $\|\cdot\|_*$, *with the subspace* E_*, *has been introduced in Lemma* 5.7.1. *Let* $\tilde{x}_n \in \Omega_n$ *denote an approximate solution to* $\mathbf{T}_n x_n = x_n$. *Suppose also that* $\mathbf{I} - \mathbf{T}_n'(\tilde{x}_n)$ *is continuously invertible in* E_n *at* $\tilde{x}_n$, *satisfying*

$$\|[\mathbf{I} - \mathbf{T}_n'(\tilde{x}_n)]^{-1}\| = \kappa_n \leq \kappa. \tag{5.90}$$

Then, for a sufficiently small meshwidth h,

$$\gamma_n \equiv (1 + \kappa_n\|\mathbf{P}_n\mathbf{T}'(\tilde{x}_n)\|)\ \|[\mathbf{T}' - \mathbf{P}_n\mathbf{T}'](\tilde{x}_n)\| + \kappa_n\|[\mathbf{T}_n - \mathbf{P}_n\mathbf{T}]^{'}(\tilde{x}_n)\| < 1. \tag{5.91}$$

In particular,

$$\begin{aligned} \|[\mathbf{I} - \mathbf{T}'(\tilde{x}_n)]^{-1}\| &\leq \kappa_n^{'} := \frac{1 + \kappa_n\|\mathbf{P}_n\mathbf{T}'(\tilde{x}_n)\|}{1 - \gamma_n}, &(5.92)\\ \|[\mathbf{I} - \mathbf{T}'(\tilde{x}_n)]^{-1}\|_* &= \kappa_n^* < \infty. &(5.93)\end{aligned}$$

It follows that $T'(\tilde{x}_n)\,|_{E_*}$ is a bounded linear operator into E_*, and, moreover, solutions, u of $[I - T'(\tilde{x}_n)]u = v, v \in E_*$, are necessarily in E_*. Finally, if $\|\tilde{x}_n - x_n\|_* \leq Ch^\theta$, where C does not depend on h, and $\kappa_n^* \leq \kappa^*$, then there exist δ_n and q_n $(\delta_n > 0;\ 0 < q_n < 1)$ such that

$$\sup_{\|x-\tilde{x}_n\|_*\leq\delta_n} \|\mathbf{T}'(x) - \mathbf{T}'(\tilde{x}_n)\| \leq \frac{q_n}{\kappa_n^{'}}, \tag{5.94}$$

$$\sup_{\|x-\tilde{x}_n\|_*\leq\delta_n} \|\mathbf{T}'(x) - \mathbf{T}'(\tilde{x}_n)\|_* \leq \frac{q_n}{\kappa_n^*}, \tag{5.95}$$

$$\|\tilde{x}_n - \mathbf{T}\tilde{x}_n\| \leq \frac{\delta_n(1 - q_n)}{\max\{\kappa_n^{'}, \kappa_n^*\}} = ch^\theta. \tag{5.96}$$

Proof. The bound on γ_n, as stated in (5.91), is proven through

$$\begin{aligned}\gamma_n &\equiv (1 + \kappa_n\|\mathbf{P}_n\mathbf{T}'(\tilde{x}_n)\|)\ \|[\mathbf{T}' - \mathbf{P}_n\mathbf{T}'](\tilde{x}_n)\| + \kappa_n\|[\mathbf{T}_n - \mathbf{P}_n\mathbf{T}]'(\tilde{x}_n)\|\\ &\leq (1 + \kappa_n\|\mathbf{P}_n\mathbf{T}'(\tilde{x}_n)\|)Ch^\theta\|\mathbf{T}'(\tilde{x}_n)\|_{H^1,H^{1+\theta}} + \kappa_n\|[\mathbf{T}_n - \mathbf{P}_n\mathbf{T}]'(\tilde{x}_n)\|\\ &\leq Ch^\theta.\end{aligned}$$

Here, the last inequality follows analogously to the bounds in Lemma 5.7.1, and C is a generic constant. The bounds (5.92–5.93) are restatements of (6.35–6.36) in the following chapter. The action on E_*, of $\mathbf{T}'(\tilde{x}_n)$, follows from the chain of lemmas contained in §5.4.

The existence of a finite δ_n and q_n $(\delta_n > 0;\ 0 < q_n < 1)$, such that (5.94) and (5.95) hold, follows because

$$\sup_{\|x-\tilde{x}_n\|_*\leq\delta_n} \|\mathbf{T}'(x) - \mathbf{T}'(\tilde{x}_n)\|_* \leq L_{\mathbf{T}'}\|x - \tilde{x}_n\|_* \leq L_{\mathbf{T}'}\delta_n,$$

so that we can choose $q_n = L_{\mathbf{T}'}\delta_n\zeta_n$, where $\zeta_n = \max\{\kappa_n^{'}, \kappa_n^*\}$. This provides the estimate for (5.95), with a similar estimate for (5.94). On the other hand, x_n is a fixed point of $\mathbf{T}_n$, and therefore

$$\begin{aligned}\|\tilde{x}_n - \mathbf{T}\tilde{x}_n\| &\leq \|\tilde{x}_n - x_n\| + \|\mathbf{T}_n x_n - \mathbf{T}x_n\| + \|\mathbf{T}x_n - \mathbf{T}\tilde{x}_n\|\\ &\leq (1 + L_{\mathbf{T}})\|\tilde{x}_n - x_n\| + C_1h^\theta\\ &\leq Ch^\theta.\end{aligned}$$

This implies that (5.96) holds, provided that

$$\delta_n = Ch^\theta \frac{\zeta_n}{(1-q_n)}.$$

The proof is concluded if we can show that the choices,

$$q_n = L_{\mathbf{T}'}\zeta_n\delta_n, \qquad \delta_n = Ch^\theta\zeta_n/(1-q_n),$$

are compatible with the requirement $q_n < 1$. Thus, set $\alpha = L_{\mathbf{T}'}\zeta_n$, $\beta = Ch^\theta\zeta_n$. We have

$$-\alpha\delta_n^2 + \delta_n - \beta = 0,$$

or,

$$\alpha\delta_n = \tfrac{1}{2}(1 \pm \sqrt{1-4\alpha\beta}).$$

Provided β is sufficiently small, so that $\alpha\beta \le \frac{1}{4}$, we may choose $\delta_n \le 1/(2\alpha)$ so that $q_n \le \frac{1}{2}$. Note that the bounds for ζ_n and γ_n show that ζ_n remains bounded as h decreases, so that the bound $\alpha\beta \le \frac{1}{4}$ does not depend on h.

5.8 Final Convergence Results

The following corollary expresses a summary of the major results in conjunction with the employed hypotheses.

Corollary 5.8.1. *Suppose that assumptions* 5.1–5.5 *of this chapter hold, and that* $N \le 3$.

- *Let* x_0 *be a fixed point of* $\mathbf{T}$. *We suppose that* 1 *is not an eigenvalue of the operator,* $\mathbf{T}'(x_0)$. *Then, there exist an index* n_0 *and a neighborhood of* x_0 *containing fixed points* x_n *of* $\mathbf{T}_n, n \ge n_0$, *satisfying*

$$\|x_0 - x_n\|_{H^1\times H^1} \le Ch^\theta, \quad \forall n. \tag{5.97}$$

  Here C *is a constant independent of* n *and* h.
- *Conversely, suppose* $\{\tilde{x}_n\}$ *is a sequence of approximate fixed points of* $\mathbf{T}_n$ *satisfying the three inequalities,* $\|\tilde{x}_n - x_n\| \le ch^\theta$, $\kappa_n^* \le \kappa^*$, *and* (5.90). *Then there exists a fixed point* x_0 *of* $\mathbf{T}$ *such that*

$$\|x_0 - \tilde{x}_n\|_{H^1\times H^1} \le Ch^\theta, \quad \forall n. \tag{5.98}$$

  Here C *is a constant independent of* n *and* h.

Proof. All norms used in the course of this proof, and the remark following it, will be H^1 norms or product norms. Estimate (5.97) follows from the conjunction of (6.15) and (5.53), with (6.16) and (5.46), respectively. This makes use of Lemma 5.7.1 in verifying the hypotheses of Theorem 6.2.1. These inequalities must be appropriately combined with the triangle inequality,

$$\|\mathbf{P}_n\mathbf{T}x_0 - \mathbf{T}_n\mathbf{P}_n x_0\| \leq \|(\mathbf{P}_n - \mathbf{I})\mathbf{T}x_0\| + \|\mathbf{T}x_0 - \mathbf{T}\mathbf{P}_n x_0\| + \|(\mathbf{T} - \mathbf{T}_n)\mathbf{P}_n x_0\|,$$

which yields order h^θ convergence, upon use of the Lipschitz continuity of $\mathbf{T}$, and the convergence properties of $\mathbf{P}_n$ and $\mathbf{T}_n$.

Inequality (5.98) follows from (6.33); here we have used Lemma 5.7.2 in verifying the hypotheses of Theorem 6.2.2. Inequality (6.33) must be supplemented by

$$\frac{\alpha_n}{1-q_n} \leq \left(\frac{\kappa_n'}{1-q_n}\right) \|\tilde{x}_n - \mathbf{T}\tilde{x}_n\| \leq \delta_n.$$

The choice, $\delta_n = Ch^\theta \zeta_n/(1-q_n)$, and the boundedness of ζ_n and $1/(1-q_n)$ are discussed in the proof of Lemma 5.7.2. The hypotheses of the corollary absorb those of Theorems 6.2.1, 6.2.2, due to the compactness of $\mathbf{T}'(x_0)$.

Remark 5.8.1. Estimates of order h^θ follow from Corollary 5.8.1 for $\|v - v_h\|$ and $\|w - w_h\|$, if $[v, w]$ and $[v_h, w_h]$ are fixed points of $\mathbf{T}$ and $\mathbf{T}_n$, respectively. Estimates for $\|U - U_h\|$ follow immediately from

$$\|\mathbf{U}(v,w) - U_h\| \leq \|\mathbf{U}(v,w) - \mathbf{U}(v_h, w_h)\| + \|\mathbf{U}(v_h, w_h) - U_h\|,$$

the Lipschitz property of $\mathbf{U}$, and (5.45), which provides the estimate for the second term on the right-hand side.

Part III

Mathematical Theory

6. Numerical Fixed Point Approximation in Banach Space

As we have seen in the preceding chapters, the drift-diffusion model of a steady-state semiconductor device is formed by a system of three coupled partial differential equations (PDEs) for which discrete and continuous maximum principles exist. This system of PDEs is solved by a solution vector of three function components. Moreover, a fixed point mapping $\mathbf{T}$ can be defined. Although the definition of $\mathbf{T}$ is not unique, and various decouplings are possible, as was rigorously analyzed in Chap. 4, it is possible to achieve complete decoupling, via gradient equations, when the recombination term satisfies monotonicity properties, or is taken to be zero. This is carried out by solving each of these PDEs for its corresponding component, and substituting these components in successive PDEs in a Gauss-Seidel iterative fashion. Fixed points of such a mapping then coincide with solutions to the drift-diffusion model. Iteration with this mapping $\mathbf{T}$ defines an algorithm for the solution of the drift-diffusion model, typically termed Gummel iteration in the literature. It is really Picard iteration for the map $\mathbf{T}$. The Lipschitz constant has been examined in detail in Chap. 4.

Algorithmic solution methods fall into two basic categories. In addition to the successive approximation just discussed, which is the older of the algorithms, a more recent algorithm consists of a damped Newton outer iteration, while the Jacobian is formed by linearization of the system of PDEs itself. Inner iteration is based on methods of discretization and numerical linear algebra. In contrast, a form of Newton's method is also discussed in this chapter at the operator level. It is related to the fixed point map, however, which permits us to retain the structural features of (linearized) Gummel iteration. In its application, it is therefore *different* from methods based upon the linearization of the system of PDEs. This represents one of the essential ideas of this book, and it is argued that this is critical in obtaining mesh independent constants in complexity estimates.

In order to analyze piecewise linear finite element discretizations, a companion approximation map is induced if the variational procedure, inherent in defining the successive gradient equations, is taken over piecewise linear, finite dimensional affine subspaces as discussed in Chap. 5. The fixed points of the companion map are clearly candidates for approximation of the fixed points of the solution map for the original system of PDEs. In this chapter, we deduce an approximation theory, described by two-sided estimates, for such a discretization procedure. The theory has been applied to piecewise linear finite element Galerkin approximations in the preceding chapter, which are characterized as approximate, or, numerical, fixed points. Our theory is based upon an operator calculus developed by Krasnosel'skii and his collaborators (cf. [90]), in which both the fixed points of the solution map are are approximated 'a priori' by fixed points of the numerical map, and also fixed points of the solution map are located in an 'a posteriori' manner near fixed points, or approximate fixed points, of the numerical map. This chapter will develop the general theory, including the linearization, especially of the numerical fixed point map. The linearization results are not included in the original theory, and are due to the author.

6.1 Linear Theory: Staircase to the Nonlinear Theory

In the theory of linear differential operators and their approximation, it has been traditional to view them as defined from smooth spaces to less smooth spaces, possibly dual spaces in the context of variational formulations. In the case of a positive definite, self adjoint mapping, it has been unnecessary to use the inverse formulation explicitly, though such formulations and related issues of smoothness enter implicitly in determining convergence rates. In variational formulations, the differential operator is realized as a specialized distribution by means of Lax-Milgram type results. This is a natural generalization of the property of self adjointness. A very powerful extension of this idea was introduced by Babuška and Aziz in [8] (see also [19]); a readable account is contained in the recent book by Brenner and Scott [17]. This theory is directed to the formulation and approximation of linear saddle point problems. Continuous bilinear forms, satisfying an inf-sup condition and some auxiliary conditions, were identified as properly defining an invertible operator framework, allowing for analysis and approximation theory. Without exploiting it directly, these authors created an underlying fixed point map, via the continuous inverse map. For the reader's benefit, we summarize here the essential features. We shall not present the most general formulation, for simplicity. In [8], it is desired to solve the operator equation, $\mathbf{L}u = f$, approximately. If B denotes the bilinear form of the weak formulation on a Hilbert space E, assume:

– continuity:

$$|B(v,w)| \leq C_1 \|v\| \, \|w\|, \tag{6.1}$$

– sup condition:

$$\text{For } w \neq 0, \ \sup_v |B(v,w)| > 0, \tag{6.2}$$

– inf-sup condition:

$$\inf_{\|v\|=1} \sup_{\|w\|\leq 1} |B(v,w)| \geq C_2 > 0. \tag{6.3}$$

Assume also sup and inf-sup conditions on an approximation space, E_n:

$$\text{For } \psi \neq 0, \ \sup_{\phi} |B(\phi,\psi)| > 0, \tag{6.4}$$

$$\inf_{\|\phi\|=1} \sup_{\|\psi\|\leq 1} |B(\phi,\psi)| \geq c_2 > 0. \tag{6.5}$$

One conclude the Galerkin approximation, u_n, is well defined by the relation,

$$B(u_n,\psi) = (\mathbf{J}f,\psi), \ \forall\psi \in E_n, \tag{6.6}$$

where $\mathbf{J}$ denotes the Riesz map. Moreover, u_n is within a metric distance,

$$\delta_*\{1 + (C_1/c_2)\}, \tag{6.7}$$

of u, where

$$\delta_* := \|u - u_*\|, \tag{6.8}$$

and u_* is arbitrary in E_n.

We note briefly the role of the hypotheses.

1. (6.1) $\Rightarrow$ $\mathbf{L}$ may be identified with a continuous linear map $\mathbf{R}$ on E.

2. (6.3) $\Rightarrow$ $\mathbf{R}^{-1}$ exists on a closed domain of E.

3. (6.2) $\Rightarrow$ Domain and range of $\mathbf{R}$ are all of E.

There is a fixed point formulation, if 1 is not a spectral value discussed below:

$$\mathbf{T}u = u, \ \ \mathbf{T}v := (\mathbf{I} - \mathbf{R})^{-1}(v - \mathbf{J}f), \ \ \mathbf{R} = \mathbf{JL}. \tag{6.9}$$

$\mathbf{T}$ is affine. Its domain independent derivative is defined by

$$\mathbf{T}'v = (\mathbf{I} - \mathbf{JL})^{-1}v, \ \forall v \in E. \tag{6.10}$$

Note that 1 is an eigenvalue of $\mathbf{T}'$ if and only if 0 is an eigenvalue of $\mathbf{L}$. The latter is excluded, so the exclusion of 1 as a spectral value is a solvability hypothesis when $\mathbf{T}$ and $\mathbf{T}'$ are compact.

In the discussion to follow, the reader may visualize the solvability conditions as generalized by the uniform invertibility condition on the derivative of the fixed point map at the fixed point. In the setting here, this invertibility is simply an eigenvalue exclusion condition, via the compactness properties. We now introduce this theory; in §6.4, it is shown to extend the inf-sup theory.

Given a fixed point x_0 of a smooth mapping $\mathbf{T}$, a numerical approximation map $\mathbf{T}_n$, and a linear projection map $\mathbf{P}_n$, a theory is constructed to estimate $\|x_n - \mathbf{P}_n x_0\|$, where $\mathbf{T}_n x_n = x_n$. In fact, the authors of [90] characterize the map $\mathbf{P}_n\mathbf{T}$ as the "Galerkin" approximate map, and $\mathbf{T}_n$ as a "perturbed Galerkin" map. Since x_0 is a fixed point of $\mathbf{T}$, the estimates represent the dispersion between these two mappings. We note that the map $\mathbf{P}_n\mathbf{T}$ is not actually implemented numerically, though it has a convergence rate which is readily estimated. It is *not* standard terminology in the numerical analysis community to refer to this as the Galerkin map. In fact, as the application of the preceding chapter has shown, one uses the Galerkin variational formulation as a starting point, and one designs the numerical fixed point map to accommodate this framework. Now, the manner in which the 'a priori' estimates are derived is to deduce a zero of the map $\mathbf{A} = \mathbf{I} - \mathbf{T}_n$, in a ball centered at $x_* = \mathbf{P}_n x_0$, by constructing an equivalent contraction map given by (6.23) below. The methodology involves derivative inversion and a mean value calculus. The result is stated as Theorem 6.2.1, and follows from the general Lemma 6.2.1. A similar approach is employed for the 'a posteriori' estimates . The relevant result is Theorem 6.2.2. We shall single out the essential hypotheses which emerge. They are stated here in somewhat more restrictive form than below, since this is the approximate format of the application procedure:

i) invertibility of $\mathbf{I} - \mathbf{T}'(x_0)$;
ii) uniform approximation of $\mathbf{T}$ by $\mathbf{T}_n$, and of $\mathbf{T}'$ by $\mathbf{T}'_n$ on appropriate bounded sets;
iii) continuity of $\mathbf{T}'$;
iv) continuity (uniform in n) of $\mathbf{T}'_n$;
v) convergence of $\mathbf{P}_n$ to $\mathbf{I}$ uniformly on appropriate bounded subsets of compactly embedded subspaces.

In our application of this theory to the semiconductor model, we shall work with energy norms, augmented by L_∞ norms when appropriate. As a model for what follows, the reader can think of a single smooth vector function p, specifying boundary values, and of an approximation p_n to p. The flat E_n may be thought of as representing the linear span of p_n and the linear space of piecewise linear trial functions, vanishing on the Dirichlet boundary. In this concrete model, $\mathbf{P}_n$ can be realized as an orthogonal projection in Hilbert space.

6.2 Nonlinear Estimation and the Operator Calculus

Thus, let E be a Banach space, and suppose $\mathbf{T}$ is a mapping from an open set Ω in E into E. We assume the existence of a fixed point x_0 for $\mathbf{T}$:

$$\mathbf{T}x_0 = x_0. \tag{6.11}$$

If $\{E_n\}$ denotes a sequence of subspaces of E of dimension $r(n) \geq n$, suppose that $\mathbf{T}_n : \Omega_n \mapsto E_n, \Omega_n \subset E_n$, has a fixed point:

$$\mathbf{T}_n x_n = x_n. \tag{6.12}$$

Finally, let $\{\mathbf{P}_n\}$ be a family of linear projections of E onto E_n. We shall describe here the framework of the operator calculus developed by Krasnosel'skii et al [90] for the convergence of the solutions of discretizations of fixed point equations (6.12) to the solutions of the original fixed point equation (6.11). First, we demonstrate in §6.2.1 that, for sufficiently large n, a solution x_n to the discretized problem (6.12) exists close to all solutions x_0 to the original problem (6.11). Second, we show in §6.2.2 that, for sufficiently large n, a solution x_0 to the original problem (6.11) exists close to all solutions x_n to the discretized problem (6.12).

6.2.1 'A Priori' Estimates and Asymptotic Linearity

We examine the degree to which (6.12) approximates (6.11) by examining the size of the operators,

$$\mathbf{R}_n = \mathbf{T}_n\mathbf{P}_n - \mathbf{P}_n\mathbf{T}, \tag{6.13}$$

defined in E.

Our first convergence result is adapted from Theorem 19.1 in [90]. It represents a generalization of that result, permitting the assumption of continuity of $\mathbf{T}'$ and $\mathbf{T}'_n$ on part or all of a subspace E_* of E with a stronger metric. Such a generalization is essential for the applications, where typically E is realized as an energy space, and E_* is a Banach subspace defined by intersection with L_∞. In these applications, $\mathbf{T}'$ is defined on all of E, but is continuous only on E_*. One still wishes the error estimates in the metric of E, but the fixed point calculations, delicately, must take into account certain invariance on subsets of E_*. The generalizations considered here, then, are not trivial ones, or unnecessary ones.

Theorem 6.2.1. *Let the operators $\mathbf{T}$ and $\mathbf{P}_n\mathbf{T}$ be Fréchet-differentiable in Ω, and $\mathbf{T}_n$ Fréchet differentiable in Ω_n. Assume that* (6.11) *has a solution $x_0 \in \Omega$ and let $\|\cdot\|_*$ be a norm, defined on a closed subspace E_*, satisfying $\|v\| \leq \|v\|_*, v \in E_*$, where $x_0 \in E_*$ and $E_n \subset E_*, \forall n$. If $\mathbf{T}(\mathbf{P}_n x_0)$ is approximated by $\mathbf{T}_n(\mathbf{P}_n x_0)$, with arbitrary precision in the norm, $\|\cdot\|_*$, as $n \to \infty$, if $\mathbf{T}$ is continuous in this norm on E_*, and if*

$$\|\mathbf{P}_n x_0 - x_0\|_* \to 0, \quad n \to \infty,$$

then

$$\|\mathbf{R}_n x_0\|_* \to 0, \quad n \to \infty. \tag{6.14}$$

Suppose also that:

1. *The linear operator* $\mathbf{I} - \mathbf{T}'(x_0)$ *is continuously invertible in* E;
2. $\mathbf{T}'$ *is continuous in* E_* *at* x_0 *with respect to the uniform operator topologies as derived from norms in both* E *and* E_*;
3. $\mathbf{P}_n\mathbf{T}'(\mathbf{P}_n x_0)$ *is approximated by* $\mathbf{T}'_n(\mathbf{P}_n x_0)$, *as* $n \to \infty$, *in the uniform operator topologies on* E_n, *as derived from both norms;*
4. $\mathbf{T}'_n(\mathbf{P}_n x_0)$ *is approximated by* $\mathbf{T}'_n(x)$, *uniformly in* x, *on a neighborhood of* $x_0 \in E_*$, *in the uniform operator topologies on* E_n *as derived from both norms.*

Then there exist n_0 *and* $\delta_0 > 0$ *such that, when* $n \geq n_0$, *equation* (6.12) *has a unique solution* x_n *in the set*

$$\mathcal{U}_n := \{x \in \Omega_n : \|x - x_0\|_* \leq \delta_0\}.$$

Moreover, in terms of the original norm, $\|\cdot\|$,

$$\|x_n - x_0\| \leq \|[\mathbf{I} - \mathbf{P}_n]x_0\| + \|x_n - \mathbf{P}_n x_0\| \to 0 \quad \text{as} \quad n \to \infty, \tag{6.15}$$

and $\|x_n - \mathbf{P}_n x_0\|$ *satisfies the following two-sided estimate* $(c_1, c_2 > 0)$:

$$c_1\|\mathbf{R}_n x_0\| \leq \|x_n - \mathbf{P}_n x_0\| \leq c_2\|\mathbf{R}_n x_0\|. \tag{6.16}$$

Remark 6.2.1. Note that in this theorem the actual rate of convergence depends only on the terms in the two-sided estimate (6.16). The additional convergence assumptions need not hold with this same rate.

Proof. The verification of (6.14) uses the continuity of $\mathbf{T}$ at x_0, as well as the approximation properties of $\mathbf{T}_n$ and $\mathbf{P}_n$. Note that hypothesis (4) allows us to assert that, for any $\epsilon > 0$, there exist n_ϵ and $\delta_\epsilon > 0$ such that

$$\|\mathbf{T}'_n(x) - \mathbf{T}'_n(\mathbf{P}_n x_0)\|_* \leq \epsilon \quad \text{for} \quad (n \geq n_\epsilon;\ \|x - \mathbf{P}_n x_0\|_* \leq \delta_\epsilon,\ x \in \Omega_n). \tag{6.17}$$

The proof now proceeds in three parts, summarized here at the outset.

- There exist constants κ and κ' such that

$$\|[\mathbf{I} - \mathbf{T}'_n(\mathbf{P}_n x_0)]^{-1}\|_* \leq \kappa, \quad \|\mathbf{I} - \mathbf{T}'_n(\mathbf{P}_n x_0)\|_* \leq \kappa', \tag{6.18}$$

  for $n \geq n_*$; here n_* is described in the statement of the theorem.
- The numbers α_n^*, for α_n^* defined by

$$\alpha_n^* = \|[\mathbf{I} - \mathbf{T}'_n(\mathbf{P}_n x_0)]^{-1}(\mathbf{I} - \mathbf{T}_n)\mathbf{P}_n x_0\|_*, \tag{6.19}$$

  satisfy

$$\frac{1}{\kappa'}\|\mathbf{R}_n(x_0)\|_* \leq \alpha_n^* \leq \kappa\|\mathbf{R}_n(x_0)\|_* \tag{6.20}$$

  for $n \geq n_*$. A similar statement holds for α_n defined by

$$\alpha_n = \|[\mathbf{I} - \mathbf{T}'_n(\mathbf{P}_n x_0)]^{-1}(\mathbf{I} - \mathbf{T}_n)\mathbf{P}_n x_0\|. \tag{6.21}$$

– The statement of the theorem concerning x_n, n_0, and δ_0 holds, and we have the bounds, expressed via the norm on E,

$$\frac{\alpha_n}{1+q} \leq \|x_n - \mathbf{P}_n x_0\| \leq \frac{\alpha_n}{1-q}, \tag{6.22}$$

for fixed $0 < q < 1$.

The inequalities (6.20) in the second part follow from routine calculations. We now consider the first part, and the associated inequalities contained in (6.18). One wishes to conclude from the existence of $[\mathbf{I} - \mathbf{T}'(x_0)]^{-1}$, the existence of $[\mathbf{I} - \mathbf{T}'_n(\mathbf{P}_n x_0)]^{-1}$. This is managed easily as follows. By the hypothesis that $\mathbf{T}'$ is continuous at $x_0 \in E_*$, we conclude by the perturbation lemma, Lemma 6.2.2, that $\mathbf{I} - \mathbf{P}_n\mathbf{T}'(\mathbf{P}_n x_0)$ is invertible on $E_n \subset E_*$, for n sufficiently large, with uniformly bounded inverses. The hypothesis on the approximation of $\mathbf{P}_n\mathbf{T}'(\mathbf{P}_n x_0)$ by $\mathbf{T}'_n(\mathbf{P}_n x_0)$ then transfers this result to the mapping $\mathbf{I} - \mathbf{T}'_n(\mathbf{P}_n x_0)$, via the perturbation lemma. This yields the first inequality of (6.18). The verification of the second inequality is even simpler, and follows from the same chain of approximations, without use of the perturbation lemma. The third part above is a consequence of Lemma 6.2.1, stated at the conclusion of the proof. The identifications,

$$\mathbf{A} = \mathbf{I} - \mathbf{T}_n, \quad x_* = \mathbf{P}_n x_0, \quad X = E_n,$$

are made. If the inequality (6.17) is employed with $\epsilon_0 = q/\kappa$, $0 < q < 1$ arbitrary, then the first hypothesis of the lemma is satisfied for some $\delta_0 \leq \delta_*$; the second hypothesis is satisfied in a similar way by deriving analogues of (6.17, 6.18) with respect to the norm on E. The third hypothesis is satisfied for n_0 sufficiently large by the second relation in (6.20). Since the bounds (6.16) of the theorem follow from the conjunction of (6.22) and the E-norm analogue of (6.20), the proof is complete. Note that, by selecting δ_0 sufficiently small, we may assume that x_n is the unique element in Ω_n within a distance of δ_0 from x_0.

The following lemma is a refinement of [90, Lemma 19.1], in the sense that an additional norm is introduced. In so doing, we distinguish between the existence of the fixed point in a ball completely determined by the stronger metric, and the estimations, carried out in the weaker metric. The proof is easily deduced from [90], and is based upon the contraction mapping,

$$\mathbf{B}x = x_* - [\mathbf{A}'(x_*)]^{-1}\{\mathbf{A}x_* + [\mathbf{A}x - \mathbf{A}x_* - \mathbf{A}'(x_*)(x - x_*)]\}, \tag{6.23}$$

where x_* is described in the lemma to follow. We cite only the more general result. The reason for proceeding in this way is to support the two major theorems of §6.2, which introduce a subspace E_*.

Lemma 6.2.1. *Let* $\mathbf{A}$ *be an operator in a Banach space* X *which is Fréchet differentiable in a closed ball centered at* x_*. *Let* X_* *be a Banach subspace* *of X, normed by $\|\cdot\|_*$, and containing x_*. It is assumed that this norm is not numerically less than $\|\cdot\|$ on X_*. Suppose $[\mathbf{A}'(x_*)]^{-1}$ exists as a bounded linear operator on X, and that the following conditions hold:*

$$\sup_{\|x-x_*\|_*\leq\delta_0} \|[\mathbf{A}'(x_*)]^{-1}[\mathbf{A}'(x)-\mathbf{A}'(x_*)]\|_* \leq q, \tag{6.24}$$

$$\sup_{\|x-x_*\|_*\leq\delta_0} \|[\mathbf{A}'(x_*)]^{-1}[\mathbf{A}'(x)-\mathbf{A}'(x_*)]\| \leq q, \tag{6.25}$$

$$\alpha^* := \|[\mathbf{A}'(x_*)]^{-1}\mathbf{A}(x_*)\|_* \leq \delta_0(1-q), \tag{6.26}$$

for some δ_0 and $0<q<1$. Then the equation $\mathbf{A}x=0$ has a unique solution x_0 in the ball, $\|x_0-x_*\|_*\leq\delta_0$. If, following (6.26), α is defined in terms of the norm on X by

$$\alpha := \|[\mathbf{A}'(x_*)]^{-1}\mathbf{A}(x_*)\|, \tag{6.27}$$

then the estimate,

$$\frac{\alpha}{1+q} \leq \|x_0-x_*\| \leq \frac{\alpha}{1-q}, \tag{6.28}$$

holds.

Remark 6.2.2. The number q here serves as a contraction constant for the mapping $\mathbf{B}$ of (6.23) on the ball $\|x_0-x_*\|_*\leq\delta_0$. The two-sided inequality, (6.16), expressed in Theorem 6.2.1, depends upon the inequality, (6.22). The number q is fixed here. It is an interesting question whether a refined version of (6.22) holds, with a sequence q_n replacing q, $q_n\to 0$. In this case, we can call the approximations defined by x_n asymptotically linear , since

$$\alpha_n \sim \|x_n-\mathbf{P}_n x_0\|, \tag{6.29}$$

with the usual meaning that the quotients tend to one. This constitutes a natural extension of the relation of equality, which holds in the affine case (cf. §6.4), and accounts for the use of the description of asymptotic linearity.

We recall that $\alpha_n\to 0$, $\alpha_n^*\to 0$, as a consequence of (6.16, 6.20), and the limit (6.14). This result will play a significant role in the sequel in allowing the replacement of the numerical fixed point by a finite number of Newton iterates, where the number does not depend on n. We have the following.

Corollary 6.2.1. *Suppose that the hypotheses of Theorem 6.2.1 hold, with item four strengthened as follows: $\mathbf{T}_n'(x)$ is Lipschitz continuous on $\mathcal{U}_n\subset E_*$, with constant $2C>0$, independent of n, in the uniform operator topologies as derived from both norms. Then the numbers q_n, defined for n such that* $\max(\alpha_n,\alpha_n^*)<\frac{1}{2\kappa C}$ *by,*

$$q_n = \frac{1}{2}\left(1-\sqrt{1-2\max(\alpha_n,\alpha_n^*)\kappa C}\right), \tag{6.30}$$

satisfy $q_n \to 0$. *Moreover, if the identifications,* $x_0 \mapsto x_n$, $x_* \mapsto \mathbf{P}_n x_0$, *and* $\delta_0 \mapsto \delta_n$ *are made, then the mapping,* $\mathbf{A} = \mathbf{I} - \mathbf{T}_n$, *satisfies the three conditions of Lemma* 6.2.1, *so that*

$$\frac{\alpha_n}{1+q_n} \le \|x_n - \mathbf{P}_n x_0\| \le \frac{\alpha_n}{1-q_n} \le \delta_n. \tag{6.31}$$

In particular, the asymptotic relation (6.29) *holds.*

Proof. Set $\epsilon_n = \frac{q_n}{\kappa}$, where κ serves as a bound in both norms. Then (6.24) and (6.25) hold, with the stated identifications. With the choice of $\delta_n = \frac{\epsilon_n}{2C}$, we have the equation,

$$\max(\alpha_n, \alpha_n^*) = \frac{q_n(1-q_n)}{2\kappa C} = \delta_n(1-q_n), \tag{6.32}$$

so that (6.26) holds, and hence (6.31), and finally, (6.29).

6.2.2 'A Posteriori' Estimates

As a converse, we assert that, for sufficiently large n, a solution x_0 to (6.11) exists near the solution x_n to the discretized problem, (6.12). As a generalization of [90, Theorem 19.2] we state the following.

Theorem 6.2.2. *As in Theorem* 6.2.1, *let* $\|\cdot\|_*$ *be a norm, defined on a closed subspace* E_* *of* E, *satisfying* $\|v\| \le \|v\|_*, v \in E_*$, *where* $E_n \subset E_*, \forall n$. *Let the operators* $\mathbf{T}$, $\mathbf{P}_n\mathbf{T}$ *and* $\mathbf{T}_n$ *be Fréchet differentiable in some neighborhood of the approximate fixed point* $\tilde{x}_n \in \Omega_n$. *Suppose* $\mathbf{I} - \mathbf{T}'_n(\tilde{x}_n)$ *is continuously invertible in* E_n,

$$\|[\mathbf{I} - \mathbf{T}'_n(\tilde{x}_n)]^{-1}\| = \kappa_n,$$

that $\mathbf{T}'(\tilde{x}_n)|_{E_*}$ *is a bounded linear operator into* E_*, *and that solutions of* $[\mathbf{I} - \mathbf{T}'(\tilde{x}_n)]u = v, v \in E_*$, *are necessarily in* E_*. *Suppose also that*

$$\gamma_n := (1+\kappa_n\|\mathbf{P}_n\mathbf{T}'(\tilde{x}_n)\|)\ \|[\mathbf{T}'-\mathbf{P}_n\mathbf{T}'](\tilde{x}_n)\|+\kappa_n\|\mathbf{T}'_n(\tilde{x}_n)-\mathbf{P}_n\mathbf{T}'(\tilde{x}_n)\| < 1.$$

Let $0 < q_n < 1$ *be given. Provided* $\mathbf{T}'$ *is continuous on* E_* *at* $\tilde{x}_n$, *with respect to the uniform operator topologies derived from both norms, and* $\tilde{x}_n$ *is sufficiently close to* $\mathbf{T}\tilde{x}_n$, *as measured by a number* δ_n, *specified in* (6.39) *below, then the equation* (6.11) *has a unique solution* x_0 *in the ball* $\|x - \tilde{x}_n\|_* \le \delta_n$, *and in terms of the norm* $\|\cdot\|$ *we have the error estimate,*

$$\frac{\alpha_n}{1+q_n} \le \|\tilde{x}_n - x_0\| \le \frac{\alpha_n}{1-q_n}, \tag{6.33}$$

where

$$\alpha_n := \|[\mathbf{I} - \mathbf{T}'(\tilde{x}_n)]^{-1}(\tilde{x}_n - \mathbf{T}\tilde{x}_n)\|. \tag{6.34}$$

Remark 6.2.3. Again the actual rate of convergence depends only on the terms in the two-sided estimate (6.33), while the additional convergence assumptions need not hold with this same rate. Also, $\tilde{x}_n = x_n$ is not required.

Proof. There are two major components of the proof:

- The invertibility of $[\mathbf{I} - \mathbf{T}'_n(\tilde{x}_n)]$ on E_n is transferred to the invertibility of $[\mathbf{I} - \mathbf{T}'(\tilde{x}_n)]$ on E; specifically,

$$\|[\mathbf{I} - \mathbf{T}'(\tilde{x}_n)]^{-1}\| \leq \kappa'_n := \frac{1 + \kappa_n \|\mathbf{P}_n \mathbf{T}'(\tilde{x}_n)\|}{1 - \gamma_n}. \tag{6.35}$$

  This invertibility guarantees a comparable bound,

$$\|[\mathbf{I} - \mathbf{T}'(\tilde{x}_n)]^{-1}\|_* \leq \kappa^*_n, \tag{6.36}$$

  on E_*.
- Lemma 6.2.1 is applied, with $\mathbf{A} = \mathbf{I} - \mathbf{T}(\tilde{x}_n)$, and $x_* = \tilde{x}_n$, to deduce the fixed point of $\mathbf{T}$, and the estimates (6.33). The conditions required for applying this lemma are specified by

$$\sup_{\|x - \tilde{x}_n\|_* \leq \delta_n} \|\mathbf{T}'(x) - \mathbf{T}'(\tilde{x}_n)\|_* \leq \frac{q_n}{\kappa^*_n}, \tag{6.37}$$

$$\sup_{\|x - \tilde{x}_n\|_* \leq \delta_n} \|\mathbf{T}'(x) - \mathbf{T}'(\tilde{x}_n)\| \leq \frac{q_n}{\kappa'_n}, \tag{6.38}$$

$$\|\tilde{x}_n - \mathbf{T}\tilde{x}_n\|_* \leq \frac{\delta_n(1 - q_n)}{\max\{\kappa^*_n, \kappa'_n\}}. \tag{6.39}$$

The proof of the estimate (6.35) proceeds in three stages, with accompanying estimates, and the first and third stages are based upon the perturbation lemma stated at the conclusion of the proof. These stages are enumerated as follows.

1. $[\mathbf{I} - \mathbf{P}_n \mathbf{T}'(\tilde{x}_n)]$ is invertible, with inverse bound, as a mapping on E_n, given by

$$\kappa_n / (1 - \kappa_n d_n),$$

   where d_n is defined by

$$d_n = \|\mathbf{P}_n \mathbf{T}'(\tilde{x}_n) - \mathbf{T}'_n(\tilde{x}_n)\|.$$

   The estimate utilizes Lemma 6.2.2, with $\|\mathbf{A}^{-1}\|$ estimated in terms of κ_n and $\|\mathbf{B}\|$ in terms of d_n.
2. The inverse bound for the map of stage 1, acting on E, is given by

$$\tau_n / (1 - \kappa_n d_n),$$

   where $\tau_n = 1 + \kappa_n \|\mathbf{P}_n \mathbf{T}'(\tilde{x}_n)\|$. This result is immediate from the representation,

$$[\mathbf{I} - \mathbf{P}_n \mathbf{T}'(\tilde{x}_n)]^{-1} = [\mathbf{I} + (\mathbf{I} - \mathbf{P}_n \mathbf{T}'(\tilde{x}_n))^{-1} \mathbf{P}_n \mathbf{T}'(\tilde{x}_n)],$$

   which allows a shift from E_n to E, and use of the result of stage 1.

3. The final bound, (6.35), is given by a routine application of Lemma 6.2.2, in conjunction with the result of stage 2 and the definition of κ_n'. The bound (6.36) now follows from the open mapping theorem of functional analysis; the continuity and surjectivity of the linear mapping, $\mathbf{I}-\mathbf{T}'(\tilde{x}_n)$, are assumed in the hypotheses of the theorem.

We now proceed to use Lemma 6.2.1. It is possible to define α_n by (6.34) and the comparable quantity α_n^*. By the inequalities (6.37–6.39), it follows that the hypotheses of Lemma 6.2.1 are satisfied. The conclusions are precisely the existence of x_0 and the bounds (6.33).

We close with the perturbation lemma used in the proofs of the theorems presented in this subsection.

Lemma 6.2.2. *Suppose that* $\mathbf{A}$ *and* $\mathbf{B}$ *are bounded linear operators on a Banach space* X *such that* $\mathbf{A}^{-1}$ *exists with* $\|\mathbf{A}^{-1}\|\;\|\mathbf{B}\| < 1$. *Then* $\mathbf{A}+\mathbf{B}$ *is invertible, and an inverse bound is given by*

$$\|[\mathbf{A}+\mathbf{B}]^{-1}\| \leq \frac{\|\mathbf{A}^{-1}\|}{1-\|\mathbf{A}^{-1}\|\;\|\mathbf{B}\|}. \tag{6.40}$$

6.3 Approximate Fixed Points via Newton's Method

It is natural to consider the extent to which the theory of the preceding section persists when the fixed points are computed by a systematic approximation procedure. We shall consider this very question in the case when Newton's method is applied to the numerical map, $\mathbf{T}_n$. What is interesting is that the hypotheses, which account for the success of the Krasnosel'skii calculus, also guarantee a corresponding replacement theory in terms of Newton's method. The reader should realize that we are presenting primarily a theoretical method here, since the linearization is based upon the fixed point map, and certain designated computations may be implicit rather than explicit in character. In other words, the procedure is a pre-algorithm, rather than an algorithm. Nonetheless, in certain applications, specifically that of this monograph, as described in the following chapter, the Newton iterate can be formed by composition maps involving sparse linear systems of equations. For the framework, we assume an identical setting as before, in terms of a Banach space E, and a continuously embedded Banach space E_*, with norm no smaller.

What is remarkable is that Newton's method is global at the level of the estimates of the previous section. More precisely, we begin by demonstrating that the Newton iterates converge q-linearly, under the hypotheses of the preceding section, in both norms. It follows that such linear convergence will eventually guarantee the residual condition required for R-quadratic convergence, when this prevails. Here, we are employing the usage of q-linear and R-quadratic as adopted in [110].

We have a preliminary lemma guaranteeing that all iterates remain in the set, $\|x - x_0\|_* \leq \delta_0$. This requires an assumption on the first iterate. For conciseness, we set $\mathbf{S}_n = \mathbf{I} - \mathbf{T}_n$.

Lemma 6.3.1. *Suppose that the bounds, $\|[\mathbf{S}_n'(\mathbf{P}_n x_0)]^{-1}\|_* \leq \kappa$, hold and that the choices, $\epsilon = q/\kappa$, δ_0, and n_0 are made in (6.17), as in the proof of Theorem 6.2.1, for fixed $0 < q < 1$. Thus, for $n \geq n_0$, let x_n be an appropriate uniquely determined fixed point of $\mathbf{T}_n$ within the set, $\mathcal{U}_n = \{x \in \Omega_n : \|x - x_0\|_* \leq \delta_0\}$. Then Newton's method is consistent, in that sequential iterates are well defined, if based upon the sequence,*

$$u_n^{k+1} - u_n^k = -[\mathbf{S}_n'(u_n^k)]^{-1}\mathbf{S}_n(u_n^k), \tag{6.41}$$

provided this sequence begins with any $u^0 \in \mathcal{U}_n$ within $*$-norm distance $\delta_0/2$ of x_0, such that the first iterate satisfies

$$\|u_n^1 - u^0\|_* \leq \frac{1}{2}(1 - q_*)\delta_0,$$

where

$$q_* := 2\frac{q}{1-q} < 1.$$

In this case, the entire Newton sequence is in $\mathcal{U}_n$.

Proof. By use of Lemma 6.2.2, and the assumed bounds of the lemma, we may conclude the existence of a uniform bound on the inverse derivative mappings given by

$$\|[\mathbf{S}_n'(x)]^{-1}\|_* \leq \frac{\kappa}{1-q}, \quad x \in \mathcal{U}_n. \tag{6.42}$$

By use of the definition of the Newton increment, (6.41), we immediately obtain the estimate,

$$\|u_n^{k+1} - u_n^k\|_* \leq \frac{\kappa}{1-q}\|\mathbf{S}_n(u_n^k)\|_*,$$

where we have used (6.42). In order to estimate the residual term, we employ the integral representation,

$$\mathbf{S}_n(u_n^k) = \tag{6.43}$$

$$\int_0^1 [\mathbf{S}_n'(u_n^{k-1} + t(u_n^k - u_n^{k-1})) - \mathbf{S}_n'(u_n^{k-1})](u_n^k - u_n^{k-1})\, dt,$$

and estimate (6.43) by use of the inequality,

$$\|\mathbf{S}_n'(x) - \mathbf{S}_n'(y)\|_* \leq \frac{2q}{\kappa},$$

which follows from the hypotheses and the triangle inequality, for x and y in $\mathcal{U}_n$. We obtain, finally,

$$\|u_n^{k+1} - u_n^k\|_* \leq \frac{2q}{1-q}\|u_n^k - u_n^{k-1}\|_*. \tag{6.44}$$

By the definition of q_*, and the repeated use of (6.44), we obtain a standard finite geometric series estimate, allowing us to conclude that each Newton iterate is in $\mathcal{U}_n$.

Corollary 6.3.1. *Suppose that the conclusion of the previous lemma is valid, that the bounds,*

$$\|[\mathbf{S}'_n(\mathbf{P}_n x_0)]^{-1}\| \leq \kappa,$$

hold, and that the choices, $\epsilon = q/\kappa$, δ_0, *and* n_0 *are made in the E-norm analogue of* (6.17), *for fixed* $0 < q < 1$. *Then Newton's method is q-linearly convergent* (cf. (6.45)), *and we have the usual linear estimate,*

$$\|x_n - u_n^k\| \leq \frac{q_*^k}{1-q_*}\|u_n^1 - u^0\|, \tag{6.45}$$

for the convergence of the Newton sequence.

Proof. By use of the same methodology as in the proof of the preceding lemma, with the E-norm substituted for the E_*-norm, we obtain, finally,

$$\|u_n^{k+1} - u_n^k\| \leq \frac{2q}{1-q}\|u_n^k - u_n^{k-1}\|. \tag{6.46}$$

By the definition of q_*, and the repeated use of (6.46), we obtain a standard Cauchy sequence estimate for $\|u_n^l - u_n^k\|$. Passage to the limit then yields (6.45). Note that here we have used the uniqueness of x_n in $\mathcal{U}_n$, and the behavior of the residuals as estimated in the course of the proof.

The above $q-$linear convergence result is valid for the norm in E_*, as well as the norm in E, by an extended argument. It was, in fact, this argument which gave us the essential condition on the location of the iterates as embodied in Lemma 6.3.1.

The R-quadratic convergence estimate is based upon the following result, which is quoted from [66]. Because of the assumption (6.49) to follow, which is verifiable in the E_*-norm directly, but not the E-norm, we must be satisfied with convergence in this norm. Because of the product term in (6.50), this estimate is sharper than those typically appearing in the literature.

Lemma 6.3.2. *Suppose that the initial residual satisfies*

$$\|\mathbf{S}_n(u^0)\| \leq \rho^{-1}, \tag{6.47}$$

where the norms in this lemma may denote $*$-norms as well. For the Newton sequence defined by (6.41), *suppose that the inequalities,*

$$\|u_n^k - u_n^{k-1}\| \leq \kappa_*\|\mathbf{S}_n(u_n^{k-1})\|, \; k \geq 1, \tag{6.48}$$

$$\|\mathbf{S}_n(u_n^k)\| \leq \frac{h\rho}{2}\|\mathbf{S}_n(u_n^{k-1})\|^2, \; k \geq 1, \tag{6.49}$$

hold for some $h \le \frac{1}{2}$. Then the convergence is described by the error estimate,

$$\|x_n - u_n^k\| \le \frac{\theta_k \kappa_*}{h\rho} \left(\prod_{j=0}^{k} \tau_j^{2^{k-j}} \right) \frac{(1-\sqrt{1-2h})^{2^k}}{2^k}. \tag{6.50}$$

Here, $\{\theta_k\}$ and $\{\tau_k\}$ are decreasing sequences bounded by 1, given explicitly by

$$\tau_k := \sqrt{1-2h} + \frac{\theta_k(1-\sqrt{1-2h})^{2^k}}{2^k}, \quad k \ge 0, \tag{6.51}$$

$$\theta_0 := 1, \tag{6.52}$$

$$\theta_{k+1} := \frac{\theta_k^2}{2^k\sqrt{1-2h} + \theta_k(1-\sqrt{1-2h})^{2^k}}, \quad k \ge 0. \tag{6.53}$$

The manner in which this result is used is similar to the structure of the proof of Corollary 6.3.1. There, the Newton increment was estimated by an inverse bound, and the residual was estimated by the integral representation (6.43). Here, the inverse estimate is already built into (6.48), and a sharpened version of (6.46) will be employed, which uses the Lipschitz continuity of the differentiated map. In order to set the stage for a detailed study, we briefly summarize the essential properties, allowing for an R-quadratically convergent Newton iteration. We require:

1. $\mathbf{S}_n'$ is Lipschitz continuous on its domain, with Lipschitz constant $L^{\mathbf{S}_n'}$.
2. The family of inverses of $\mathbf{S}_n'$ is uniformly bounded in norm, say, by κ_*.
3. The initial residual, $\mathbf{S}_n(u^0)$, does not exceed in norm the quantity, $[2L^{\mathbf{S}_n'}\kappa_*^2]^{-1}$.

In the theorem to follow, the norm for R-quadratic convergence will be taken to be the norm in E_*, since the conditions just cited are sufficient exactly when there is norm compatibility between domain topology and the uniform operator topology. The derivation of the convergence result under these hypotheses is described in [66, §2], with slight changes in the quotation of the result, as embodied in Lemma 6.3.2, due to the fact that an approximate Newton method is described in this reference. Here, an exact Newton method for $\mathbf{T}_n$ is analyzed, which indirectly, via the Krasnosel'skii framework, translates into an approximate Newton method for $\mathbf{T}$. In the items listed above, the reader will find a striking parallel to the previous section; what may appear to be missing there is a condition guaranteeing the sufficiently small residual required for R-quadratic convergence.

Theorem 6.3.1. *Define*

$$h := \frac{L^{\mathbf{S}_n'}\kappa_*^2}{\rho} \tag{6.54}$$

and suppose that $h \le \frac{1}{2}$. Suppose that all norm identifications to follow in this theorem refer to the $*$-norm. Under the hypothesis (6.47), it follows that the *conditions* (6.48) *and* (6.49) *hold. In particular, the R-quadratic convergence estimate* (6.50) *holds, with* $h\rho$ *as defined in* (6.54) *above.*

Proof. Since (6.48) is immediate, it remains to verify (6.49), with the choice of $h\rho$ as defined in (6.54). For this end, we use the representation (6.43) as in the proof of Lemma 6.3.1. By use of the full Lipschitz continuity of $\mathbf{S}'_n$, we are able to deduce the stronger result (6.49).

We have now developed a tight linearization theory . We summarize the essential features as follows.

- Newton's method, based upon the approximate fixed point map $\mathbf{T}_n$, is globally convergent at the level of the Krasnosel'skii calculus, developed in the previous section. In particular, there is a systematic procedure for determining x_n approximately.
- The convergence is at least linear, as described in Corollary 6.3.1. The switching point to quadratic convergence takes place no later than when the residual condition (6.47) is met. When this occurs, as measured by (6.54), the convergence is described by Theorem 6.3.1.
- Only as many Newton iterates are required to approximate x_n as matches the approximation estimate for the dispersion between $\mathbf{P}_n x_0$ and x_n, as given in (6.16).
- Although it is not demonstrated above (cf. Proposition 6.3.1 to follow for details), it is possible to employ *a bounded number* of Newton iterates, with bound independent of n, while maintaining consistency with the estimates. This result depends upon asymptotic linearity, developed in Corollary 6.2.1, wherein the number q of (6.22) is replaced by a sequence $q_n \to 0$. The existence of such a bound for the number of Newton iterations, equivalent to the numerical fixed point, is a critical component of the theory (see §7.1.1).

Proposition 6.3.1. *Without loss of generality, suppose that the sequence* $\{q_n\}$, *defined in Corollary* 6.2.1, *is monotone decreasing. Suppose also that the conclusions of this corollary and Lemma* 6.3.1 *hold, under the assumption that the starting choice* u^0 *at the nth. stage is the accepted approximate fixed point* $\tilde{x}_{n-1}$ *at the* $(n-1)$*th. stage. In particular, the Newton iterates are well defined in* $\mathcal{U}_n$. *Finally, suppose that the bounds,*

$$\|[\mathbf{S}'_n(\mathbf{P}_n x_0)]^{-1}\| \le \kappa,$$

hold, and that the choices, $\epsilon = q_n/\kappa$, δ_0, *and* n_0 *are made in the E-norm analogue of* (6.17). *Then Newton's method is q-linearly convergent:*

$$\|x_n - u_n^k\| \le \frac{q_{*n}^k}{1 - q_{*n}} \|u_n^1 - u^0\|. \tag{6.55}$$

Here, we have set $q_{*n} = \frac{2q_n}{1-q_n}$. *In particular, we may select an integer* k_0, *not depending on* n, *such that*

$$\|x_n - u_n^k\| \le \delta_n := \frac{q_n}{2\kappa C}, \; k \ge k_0, \tag{6.56}$$

and we may define, $\tilde{x}_n = u_n^{k_0}$. *The error estimate,* $\| \tilde{x}_n - \mathbf{P}_n x_0\| \le 2\delta_n$, *holds. We may choose* k_0 *to be the smallest integer* k *satisfying,*

$$2\delta_0 \frac{q_{*n}^k}{1 - q_{*n}} \le \delta_n. \tag{6.57}$$

6.4 The Inf-Sup Theory As a Special Case

In this section, we shall show that the hypotheses of the Babuška-Aziz theory imply the hypotheses, and hence the conclusions, of the Krasnosel'skii theory in the case where only one norm is employed. The conclusions of this theory are seen to imply those of the inf-sup theory. These results were first published in [73]. We begin by itemizing key properties of the construction of [8]. It is now understood that E is a Hilbert space, and we identify E and E_*. In applications, E is typically the Sobolev space, H^1.

- The fixed point map, $\mathbf{T}$, is defined by (6.9). The mapping $\mathbf{R}$ in the definition of $\mathbf{T}$ is determined by the relation,

$$B(u, v) = (\mathbf{R}u, v), \quad \forall u, v \in E. \tag{6.58}$$

- If the numerical fixed point map on E_n is defined by

$$\mathbf{T}_n v = (\mathbf{I} - \mathbf{P}_n\mathbf{R})^{-1}(v - \mathbf{P}_n\mathbf{J}f), \; v \in E_n, \tag{6.59}$$

  it follows that the Galerkin approximation, written as u_n, is characterized as a fixed point of $\mathbf{T}_n$. Indeed, it is shown in ([8, pp. 187-188]) that $\mathbf{P}_n\mathbf{R}$ maps E_n into itself according to

$$B(\phi, \psi) = (\mathbf{P}_n\mathbf{R}\phi, \psi), \quad \forall \phi, \psi \in E_n. \tag{6.60}$$

  The fixed point property then follows immediately from (6.6) and (6.60).
- The domain independent derivative of $\mathbf{T}_n$ is given by

$$\mathbf{T}_n' v = (\mathbf{I} - \mathbf{P}_n\mathbf{R})^{-1} v, \quad \forall v \in E_n. \tag{6.61}$$

- If v is an eigenvector of $\mathbf{T}'$, given by (6.10), corresponding to eigenvalue, 1, then v is also an eigenvector of $\mathbf{R}$, corresponding to eigenvalue, 0. This follows from the relation,

$$\mathbf{T}' v = (\mathbf{I} - \mathbf{R})^{-1} v = v.$$

  Since the latter contradicts both the sup condition (6.2) and the inf-sup condition (6.3), 1 is not an eigenvalue. This represents the principal nonsingularity hypothesis of §6.2.

Theorem 6.4.1. *Let u denote the unique solution of the operator equation, $\mathbf{L}u = f$. Set*

$$v_n = [\mathbf{I} - \mathbf{T}_n']^{-1}(\mathbf{I} - \mathbf{T}_n)\mathbf{P}_n u. \tag{6.62}$$

Then, by definition,

$$\alpha_n = \|v_n\|, \tag{6.63}$$

and, in this affine case with $q = 0$, the fundamental result (6.22) *of the calculus of* §6.2.1 *yields the result that*

$$\alpha_n = \|u_n - \mathbf{P}_n u\|. \tag{6.64}$$

Independently, the relations,

$$\mathbf{P}_n\mathbf{R}v_n = \mathbf{P}_n\mathbf{R}\mathbf{P}_n u - \mathbf{P}_n\mathbf{J}f, \tag{6.65}$$

$$\mathbf{P}_n\mathbf{R}u = \mathbf{P}_n\mathbf{R}u_n = \mathbf{P}_n\mathbf{J}f, \tag{6.66}$$

hold. In particular, if $(\mathbf{P}_n\mathbf{R})^{-1}$ denotes inversion on E_n, it follows that yet another characterization is given by

$$\alpha_n = \|(\mathbf{P}_n\mathbf{R})^{-1}[\mathbf{P}_n\mathbf{R}(\mathbf{P}_n u - u)]\|. \tag{6.67}$$

We may infer, therefore, the inequalities,

$$\|u_n - \mathbf{P}_n u\| \le \frac{C_1}{c_2}\|\mathbf{P}_n u - u\|, \tag{6.68}$$

$$\|u_n - \mathbf{P}_n u\| \ge \frac{C_2}{C_1}\|\mathbf{P}_n u - u\|. \tag{6.69}$$

The inf-sup estimate, (6.7), *follows directly from* (6.68).

Proof. We begin with (6.62), apply the mapping, $[\mathbf{I} - \mathbf{T}_n']$, insert the representations given by (6.59) and (6.61), and simplify to obtain (6.65).

In order to obtain the second equality in (6.66), begin with the relation, $\mathbf{T}_n u_n = u_n$, substitute (6.59), and simplify. In a similar manner, the relation, $\mathbf{T}u = u$, implies

$$\mathbf{R}u = \mathbf{J}f,$$

so that both equalities in (6.66) are seen to hold. The identity (6.67) is now a simple consequence of (6.65, 6.66), and the definition (6.63) of α_n. Inequalities (6.68, 6.69) follow from standard mapping properties of the inf-sup theory, as applied to $\mathbf{R}$ and $\mathbf{P}_n\mathbf{R}$.

Remark 6.4.1. This completes the verification that the inf-sup theory is implied by the Krasnosel'skii operator calculus, at least in the Galerkin case. There are, of course, extensions of the inf-sup theory to the Petrov-Galerkin case, which is not considered here.

Remark 6.4.2. The author announced the result of Theorem 6.4.1 at the Lancaster Conference (see [72]), and published it in the memorial volume [73]. This was most fitting, since it was Farouk Odeh who encouraged T. Kerkhoven and the author to consider the Krasnosel'skii operator calculus as an analytical tool for the drift-diffusion model. Odeh's suggestion occurred just prior to the formulation in 1987 of an interesting problem by Ivo Babuška, who asked whether linear iterations for nonlinear problems were predictable, in the sense that the dictates of accuracy were sufficient to estimate the number of corresponding linear iterates. While Kerkhoven and the author did apply the nonlinear operator calculus in [74], it was not until the idea of asymptotic linearity was introduced in [73] that the problem raised by Babuška was resolved in principle. The surprising resolution is that the number of linear iterations involved in the application of Newton's method (to $\mathbf{T}_n$) does not depend on n, if the linear algebra is done exactly. The situation of approximate linear algebra is discussed in the following chapter.

7. Construction of the Discrete Approximation Sequence

Since the fundamental paper of Moser (cf. [105]), it has been understood analytically that regularization is necessary as a postconditioning step in the application of approximate Newton methods, based upon the system differential map. A development of these ideas in terms of current numerical methods and complexity estimates was given by the author in [65]. The approach of Moser is often termed Nash-Moser iteration, because of the fundamental link to generalized implicit function theorems (cf. [108]), specifically, the Nash implicit function theorem. It was proposed by the author in [70], and analyzed further in [71], to use the fixed point map as a basis for the linearization, and thereby avoid the loss of derivatives phenomenon identified by Moser, and termed a numerical loss of derivatives in [65]. In the context of numerical analysis, this loss occurs because the approximation of the identity condition, involved in approximate Newton methods, is not robust with respect to differentiation up to the order of the nonlinear differential system (see (7.8) below). In this chapter, we shall discuss the implications of this fact. It is at the core of preferring the fixed point formulation to the differential formulation. We begin with the former. Our discussion of this case is brief, because of the developments of the preceding chapters.

7.1 The Fixed Point Map as Smoother

In the preceding chapter, the Krasnosel'skii operator calculus was merged with Newton's method, for the computation of the approximate fixed points, in such a way that the approximation order was preserved. Note that an exact Newton method was introduced for the numerical fixed point map; this may be viewed as an approximate Newton method for the fixed point map itself. When interpreted in terms of the specific model of Chap. 5, this means that the order is preserved with mesh independent constants .

Definition 7.1.1. *For fixed $n \geq 1$, set $u_n^0 = \tilde{x}_{n-1}$, the accepted approximate fixed point at the previous stage. The sequence $\{u_n^k\}$ is constructed from the iterative solution of the linear system,*

$$-\mathbf{S}_n'(u_n^k)[u_n^{k+1} - u_n^k] = \mathbf{S}_n(u_n^k), \quad k \geq 0. \tag{7.1}$$

where $\mathbf{S}_n = \mathbf{I} - \mathbf{T}_n$. *The left hand side of the sequence* (7.1) *can be computed from* (5.66), *with the components of the Newton increment,* σ_n^{k+1} *and* τ_n^{k+1}, *undetermined linear combinations of basis functions, followed by use of* (5.79), *and the corresponding system for the second component. This makes direct use of the chain rule for the composition mapping, and suppresses the truncation map in the situation where maximum principles hold. From this discussion, we see that the first component of the left hand side of* (7.1) *is given by the expression,*

$$-\,[\mathbf{I} - \mathbf{V}_h'(\mathbf{U}_h(u_n^k)) \circ \mathbf{U}_h'(u_n^k)](\sigma_n^{k+1}, \tau_n^{k+1}), \tag{7.2}$$

where

$$[\sigma_n^{k+1}, \tau_n^{k+1}] = u_n^{k+1} - u_n^k. \tag{7.3}$$

The right hand side of (7.1) *is assembled through the use of the nonlinear Gauss-Seidel procedure, used to define the numerical fixed point map. Thus, the system can be finally solved by simple linear "inversion". This fixes the values, then, of the basis coefficients for* σ_n^{k+1} *and* τ_n^{k+1}. *Thus, the successive solutions induced by the compositions, when viewed as a symbolic calculation, are defined by sparse matrix calculations. These calculations are terminated with* $k = k_0$, *defined by* (6.57). *Thus, there is a bound, independent of* n, *on the number of linear problems which must be solved to obtain the desired accuracy at the* n*-th. stage. The inductive process is terminated by setting* $\tilde{x}_n = u_{k_0}^n$.

This is one way of resolving the smoothing dilemma; the fixed point map is inherently smoothing. No loss of derivatives occurs in this case.

Remark 7.1.1. The discussion just completed does not address the issue of the numerical linear algebra required to carry out the computation. Thus, this does not represent an easily verifiable complexity estimate, in terms of an operation count, $p(n)$. Toward this end, this system has been investigated in great detail by Kerkhoven and Saad in [85] by use of GMRES (and other methods as well). GMRES (generalized minimum residual) is based upon Arnoldi's process, and was introduced in [118]. The usage here may be thought of as a way of approximately implementing the solution of (7.1), which defines the exact Newton sequence for the numerical fixed point. In fact, the Krylov subspace of dimension m, employed at the n-th stage, is

$$K_m = \{v_1, \mathbf{J}_n v_1, \ldots, \mathbf{J}_n^{m-1} v_1\},$$

where $\mathbf{J}_n$ is the derivative (Jacobian) of the identity shifted $\mathbf{T}_n$, and v_1 is a normalized residual. Kerkhoven and Saad are able to implement this algorithm in a Jacobian free manner, i.e. , using only values generated by $\mathbf{T}$, and derive an essential *superlinear* convergence estimate for the residual; this suggests, though not with the conclusiveness of proof, that the work function of a single GMRES iteration, depending on n, when multiplied by a fixed number of inner iterations, yields an estimate for the computational complexity of a single outer iteration, of which there are a bounded number, independent of n. Some features distinguish the present setting from that of [85]:

1. **T** is defined differently here;
2. Reference [85] is concerned with the acceleration of the fixed point iteration;
3. Outer and inner iterations are consolidated in [85].

Nonetheless, this approach complements the present study, and adheres to the philosophical principle that the iterative strategy contains its own preconditioning (see also [25] for this idea in a different context).

7.1.1 Solution of the Central Approximation Problem

In 1936, Kolmogorov [89] introduced the concept of n-dimensional diameter, or n-width, for a set $\mathcal{K}$ in a normed linear space E:

$$d_n(\mathcal{K}) = \inf_{\dim \mathcal{M}=n} \sup_{f\in\mathcal{K}} \inf_{g\in\mathcal{M}} \|f-g\|. \tag{7.4}$$

In his pioneering paper, Kolmogorov characterized the diameters of a particularly significant class for differential equations,

$$\mathcal{K} = \{f \in H^k(a,b) : |f|_k = \left\{\int_a^b |f^{(k)}(x)|^2\, dx\right\}^{1/2} \leq 1\}, \tag{7.5}$$

as viewed in the metric of $L_2(a,b)$. The eigenvalues of a natural Sturm-Liouville problem characterize the diameters; this leads directly to the exact estimate of the asymptotic order, $O(n^{-k})$. The extension of the Kolmogorov results to classes of functions of several variables was carried out in the author's doctoral thesis [61], and, in Euclidean space of dimension N, the order is $O(n^{-k/N})$. Obvious extensions give an order of $O(n^{-(k-1)/N})$ if the H^1 metric is used; here, k may also be nonintegral. For a more complete description of n-width results, the reader should consult the books [64, 98, 111].

Definition 7.1.2 (central approximation problem). *Write the system* (5.1–5.3), *as* $\mathbf{F}(z) = 0$, *and suppose the Sobolev regularity exponent of* z *is* $1+\theta$, $0 < \theta \leq 1$, *and 'a priori' bounds locate the solution* z *in a ball of radius* ρ *in* $\prod H^{1+\theta}$. *As viewed in* $\prod H^1$, *call this set* Ω. *We shall require an approximation sequence* $\{x_n\}$ *for* z *to satisfy:*

1. *Optimal Order Approximation;*
$$\|z - x_n\| \leq C d_n,\ n \to \infty.\ (d_n \approx n^{-\frac{\theta}{N}}) \tag{7.6}$$
2. *Computability;* $\forall n$, x_n *is defined by a sequence of linear systems, bounded independently of* n, *each member of which can be constructed by a (bounded) number of sparse matrix calculations.*
3. *Stability (Discrete Maximum Principle);* $\{x_n\}$ *is bounded in* $\prod L_\infty$.

The determination of such a sequence is what we have termed the central problem for drift-diffusion systems. Under the hypotheses of Chap. 5, we have given a solution of this problem (see the earlier part of this section). That solution requires the finite dimensional approximations to be computed iteratively from (7.1). The rate of convergence is the same as that of the Galerkin sequence. Note that this order $O(h^\theta)$ corresponds to $O(n^{-\frac{\theta}{N}})$ for quasi-uniform triangulations. That the resolution of the central approximation problem is quite subtle, and has bypassed the numerical loss of derivatives, is confirmed by the following section, which is included for completeness, although it is not logically essential for the results presented in this monograph. A generic situation is considered in what follows. We indicate how one would necessarily proceed if the differential map were used to define Newton's method, as in traditional approaches to this computational problem.

7.2 Smoothing for Newton Iteration: Differential Maps

Suppose that $\mathbf{F} := \Phi(D^\alpha) : B_{\eta,0} \subset Y_0 \to X_0$ is a differential mapping of order m, from $B_{\eta,0} = \{v \in Y_0 : \|v - v_0\|_{Y_0} \leq \eta\}$ to X_0, where X_0 and Y_0 are Banach function spaces. The real Hölder spaces are the prototypical examples. We make the following assumptions initially. We shall assume that $\mathbf{F}$ is continuously Lipschitz (Fréchet) differentiable on an open set U containing $B_{\eta,0}$:

$$\|\mathbf{F}'(v) - \mathbf{F}'(w)\|_{Y_0,X_0} \leq 2M\|v - w\|_{Y_0}, \qquad v, w \in B_{\eta,0}, \tag{7.7a}$$

$$\|\mathbf{F}'(v)\|_{Y_0,X_0} \leq M, \qquad v \in B_{\eta,0}. \tag{7.7b}$$

It is desired to determine a root of the operator equation, $\mathbf{F}(u) = 0$, by a Newton iterative method, involving an approximate inverse, $\mathbf{G}_h(v)$, of the map, $\mathbf{F}'(v)$, which is defined by a standard numerical method. If $\mathbf{G}(v)$ represents the actual inverse, and w the current residual, then we note that each differentiation of

$$[\mathbf{G}(v) - \mathbf{G}_h(v)]w$$

leads to a loss of order one in the convergence order, in general. In particular, the differentiation of the approximation of the identity,

$$[\mathbf{F}'(v)\mathbf{G}_h(v) - I]w, \tag{7.8}$$

is of order $O(1)$, and thus experiences a numerical loss of derivatives. Classical theories, such as the Kantorovich theory, detailed in [66], are ill equipped to deal with this situation, since they are ultimately based on use of an approximation of the identity, which, when differentiated, is of the order of a polynomial function of the grid size, chosen adaptively according to the current residual. Excluding exceptional situations, such as those of superconvergence, we conclude that there is no *general* way to implement an approximate Newton method for a differential map, defined via a numerical discretization, without further conditioning, if constants are to be mesh independent.

7.2.1 Framework for the Postconditioning Iteration

Definition 7.2.1 (scales). *We are led to introduce a scale of Banach spaces, X_σ and Y_σ, where the following properties hold:*

- *These spaces are continuously embedded, so that, for $\sigma' < \sigma$,*

$$X_0 \supset X_{\sigma'} \supset X_\sigma \supset X_\infty := \cap_{\sigma>0} X_\sigma. \tag{7.9}$$

There is a similar statement for $\{Y_\sigma\}$.

- *There exists a smoothing $\mathbf{S}_t$: $X_0 \to X_\infty$ for $t \geq 1$:*

$$\|\mathbf{S}_t v - v\|_{X_0} \to 0, \qquad \text{as } t \to \infty; \tag{7.10a}$$

$$\|\mathbf{S}_t v\|_{X_p} \leq C_{rp} t^{p-r} \|v\|_{X_r}, \qquad 0 \leq r \leq p; \tag{7.10b}$$

$$\|\mathbf{S}_t v - v\|_{X_r} \leq C_{rp} t^{-(p-r)} \|v\|_{X_p}, \qquad 0 \leq r \leq p. \tag{7.10c}$$

There is a similar set of inequalities assumed for the spaces $\{Y_\sigma\}$.

- *Hence* (cf. [108, *Lemma* 6.3.2]), *norm interpolation is possible:*

$$\|v\|_{X_r} \leq C(q, r_1, r_2) \|v\|_{X_{r_1}}^{1-q} \|v\|_{X_{r_2}}^{q}, \tag{7.11}$$

$0 \leq q \leq 1$, $0 \leq r_1 \leq r_2$, $r = (1-q)r_1 + qr_2$. *Similar interpolation is assumed to hold for the spaces, $\{Y_\sigma\}$.*

It is necessary, finally, to express how the mapping $\mathbf{F}$, its derivatives, and approximate inverses behave relative to this scale of spaces.

Definition 7.2.2. *We explicitly assume the following.*

1. $\mathbf{F} : B_{\eta,0} \cap Y_\sigma \to X_\sigma$ *is a well-defined map for $0 \leq \sigma \leq s$ and s sufficiently large* (cf. (7.20c)). *This includes the assumption that $\mathbf{F} = \Phi(D^\alpha)$, with Φ as in* (7.12) *to follow, for $a = s$.*
2. *Functional substitution is essentially of linear growth:*

$$\|\Phi(D^\alpha v)\|_{X_a} \leq C(a, \gamma, m, \Phi, v_0)(\|v - v_0\|_{Y_a} + \|v - v_0\|_{Y_\gamma}^{a} + 1) \tag{7.12}$$

for any fixed v_0, $a \geq 1$, $1 \leq \gamma \leq a$ and $\{\alpha : |\alpha| \leq m\}$. Here, Φ is sufficiently smooth on a fixed compact, convex set containing the range of the vector $D^\alpha v$.

3. *There exist numerical inversion operators with loss of derivatives. Specifically, there is a constant $\gamma \geq 1$ and a family $\mathbf{G}_h(v) : X_\gamma \to Y_0$, of linear mappings depending on parameters h and v, and a continuous monotone increasing function $\tau : [0, b] \to [0, \infty)$, $\tau(0) = 0$, such that, $\forall w \in X_\gamma$,*

$$\|[\mathbf{F}'(v)\mathbf{G}_h(v) - I]w\|_{X_0} \leq \tau(h)\|w\|_{X_\gamma}, \qquad \forall v \in B_{1,\gamma} \subset Y_\gamma. \tag{7.13}$$

Here, we have used the symbol, $B_{1,\gamma} \subset Y_\gamma$, to denote the ball of radius one.

4. *The maps* $\{\mathbf{G}_h(v)\}$ *are uniformly bounded in* h *and* v *from* X_σ *to* $Y_{\sigma-\gamma}$ *for* $\gamma \leq \sigma \leq s$:

$$\|\mathbf{G}_h(v)w\|_{Y_{\sigma-\gamma}} \leq M\|w\|_{X_\sigma}. \tag{7.14}$$

Remark 7.2.1. In this remark, we discuss the approximate Newton method with *postconditioning.* We also discuss the fundamental parameters. We begin with u_0, the initial iterate. For ease of notation, since computations proceed with reference to u_0, we identify u_0 with the center v_0 of $B_{\eta,0}$. More generally, u_0 will be the center of the closed balls $B_{c,\sigma} \subset Y_\sigma$ of radius c. We assume

$$u_0 \in Y_s, \qquad f_0 = \mathbf{F}(u_0) \in X_s, \qquad \|f_0\|_{X_0} \leq \rho^{-\lambda}, \tag{7.15}$$

where $\rho > 0$ and $1 < \lambda < s$. We select a superlinear convergence parameter, β, satisfying $1 < \beta < 2$. In terms of these quantities, we require:

$$\|f_0\|_{X_s} \leq M\rho^{(s-\lambda)\beta}. \tag{7.16}$$

We select an acceleration parameter, $\theta > 1$, and define the smoothing speeds by

$$t_k = \rho^{\theta\beta^k}, \qquad k \geq 1. \tag{7.17}$$

We assume, inductively, that $u_{k-1} \in Y_s \cap B_{1,\gamma}$ has been defined for $k \geq 1$. Then u_k is defined by Newton/postconditioning iteration:

$$u_k - u_{k-1} = -\mathbf{S}_{t_k}\mathbf{G}_{h_k}(u_{k-1})\mathbf{F}(u_{k-1}). \tag{7.18}$$

Here, the parameter h_k is subject to specification (cf. (7.22) below), so that $\tau(h_k)$ is of the order of $\mathbf{F}(u_{k-1})$. The iteration is required to converge in Y_μ for specified $\mu \geq 1$. The relations among γ, s, λ, β, θ, and μ are given in Definition 7.2.3 to follow. For ease of estimation, we assume that M is chosen so that $M \geq 1$ and

$$M \geq \sup(C_{r,p},\ C(q, r_1, r_2),\ 3C(s, \gamma, m, \Phi, u_0)), \tag{7.19}$$

where the constants appear in (7.10–7.12). The supremum is understood to be taken over r, r_1, r_2, $p \leq s + m$, and $0 \leq q \leq 1$.

Definition 7.2.3 (parametric relations). *The relations satisfied by the exponents and parameters are presented here. We require:*

$$1 < \beta < 2, \qquad 1 < \theta, \qquad 1 \leq \gamma \leq \mu < \lambda < s, \tag{7.20a}$$

$$\lambda \geq \max\{2\beta\gamma(2-\beta)^{-1},\ \beta(\gamma + \mu\beta)\} + \lambda_0(2-\beta)^{-1}\}, \tag{7.20b}$$

$$s \geq \max\{\theta\gamma(\theta-1)^{-1},\ \lambda + \theta\gamma(\beta-1)^{-1}\} + s_0 \max\{(\theta-1)^{-1},\ (\beta-1)^{-1}\}, \tag{7.20c}$$

for arbitrary $\lambda_0 > 0$, $s_0 > 0$. *The number* ρ *is required to satisfy*

$$\rho^{\ln\beta} \geq e, \tag{7.21a}$$
$$M^9\rho^{-\lambda_0} \leq 1/4, \tag{7.21b}$$
$$M^3\rho^{-s_0} \leq 1/2, \tag{7.21c}$$
$$M^4\rho^{-(\lambda-\beta\gamma)} \leq \eta/2, \tag{7.21d}$$
$$\rho^{\lambda_0\ln\beta} \geq e. \tag{7.21e}$$

The numbers h_k are selected so that

$$\tau(h_k) \leq M^5\|\mathbf{F}(u_{k-1})\|_{X_\gamma}, \qquad k = 1, \ldots. \tag{7.22}$$

7.2.2 The Superlinear Convergence Theorem

Theorem 7.2.1. *Suppose the hypotheses of Definitions* 7.2.1–7.2.3, *are satisfied, and that the initial iterate $u_0 \in B_{\eta,0} \cap Y_s$ satisfies* (7.15, 7.16). *Then the Newton iterates $\{u_k\}$, defined by the adaptive, postconditioning procedure* (7.18), *converge to a root u of $\mathbf{F}$ in $B_{\eta,0} \cap B_{1,\mu}$. The superlinear convergence in Y_0 is described by the estimate*

$$\|u - u_k\|_{Y_0} \leq M^4\rho^{-(\lambda-\beta\gamma)\beta^{k-1}}/\beta^{k-1}, \tag{7.23}$$

for $k = 1, 2, \ldots$. The superlinear convergence in Y_μ is described by

$$\|u - u_k\|_{Y_\mu} \leq M^5\rho^{-\lambda_0\beta^{k-1}}/\beta^{k-1}, \tag{7.24}$$

for $k = 1, 2, \ldots$.

Remark 7.2.2. The choices

$$\theta = 1.16, \quad \beta = 1.4, \quad \mu = \gamma, \quad \lambda = \tfrac{29}{6}\gamma, \quad s = \tfrac{47}{6}\gamma, \tag{7.25}$$

satisfy (7.20a-c), for $\lambda_0 = 0.1$ and $s_0 = 0.016\gamma$. These are not the best choices, but do reflect realistic magnitudes for the parameters. Some independent experimental calculations suggest that $\theta = 1.1435$, $\lambda = 3.2952$, $s = 7.9702$ and $\beta = 1.2446$, for $\mu = \gamma = 1$, might serve as first approximations for these parameters if the decrease of λ and s is preferred to the increase of β.

Proof. We establish the following statements recursively for $k \geq 0$:

$$\|\mathbf{F}(u_k)\|_{X_0} \leq \rho^{-\lambda\beta^k}, \tag{7.26a}$$
$$\|\mathbf{F}(u_k)\|_{X_s} \leq M\rho^{(s-\lambda)\beta^{k+1}}, \tag{7.26b}$$
$$\|\mathbf{F}(u_k)\|_{X_\gamma} \leq M^2\rho^{-(\lambda-\beta\gamma)\beta^k}, \tag{7.26c}$$
$$\|u_{k+1} - u_k\|_{Y_s} \leq \tfrac{1}{2}\rho^{(s-\lambda)\beta^{k+2}}, \tag{7.26d}$$
$$\|u_{k+1} - u_0\|_{Y_s} \leq \rho^{(s-\lambda)\beta^{k+2}}, \tag{7.26e}$$
$$\|u_{k+1} - u_k\|_{Y_0} \leq M^4\rho^{-(\lambda-\beta\gamma)\beta^k}, \tag{7.26f}$$

$$\begin{aligned}
\|u_{k+1} - u_0\|_{Y_0} &\leq \eta, && (7.26g)\\
\|u_{k+1} - u_k\|_{Y_\delta} &\leq M^5 \rho^{-\lambda_0 \beta^k}, \qquad \gamma \leq \delta \leq \mu, && (7.26h)\\
\|u_{k+1} - u_0\|_{Y_\delta} &\leq 1, \qquad \gamma \leq \delta \leq \mu. && (7.26i)
\end{aligned}$$

Suppose, then, that (7.26a–i) hold for $k < j$, where $j \geq 0$ is an arbitrary fixed integer. To verify (7.26a), for $k = j$, we write, for $j > 0$,

$$\begin{aligned}
\mathbf{F}(u_j) = &-[\mathbf{F}'(u_{j-1})\mathbf{G}_{h_j}(u_{j-1}) - \mathbf{I}]\mathbf{F}(u_{j-1})\\
&+R(u_{j-1}, u_j) + \mathbf{F}'(u_{j-1})[\mathbf{I} - \mathbf{S}_{t_j}]\mathbf{G}_{h_j}(u_{j-1})\mathbf{F}(u_{j-1}), \qquad (7.27a)
\end{aligned}$$

where

$$R(u_{j-1}, u_j) = \mathbf{F}(u_j) - \mathbf{F}(u_{j-1}) - \mathbf{F}'(u_{j-1})(u_j - u_{j-1}). \qquad (7.27b)$$

Now an application of (7.13), with $\tau(h)$ selected according to (7.22), yields

$$\|[\mathbf{F}'(u_{j-1})\mathbf{G}_{h_j}(u_{j-1}) - \mathbf{I}]\mathbf{F}(u_{j-1})\|_{X_0} \leq M^5 \|\mathbf{F}(u_{j-1})\|^2_{X_\gamma},$$

and (7.7a) gives

$$\|R(u_{j-1}, u_j)\|_{X_0} \leq M \|u_j - u_{j-1}\|^2_{Y_0},$$

via the representation for $\|R(u_{j-1}, u_j)\|_{X_0}$, specified by

$$\left\| \int_0^1 [\mathbf{F}'(u_{j-1} + r(u_j - u_{j-1})) - \mathbf{F}'(u_{j-1})](u_j - u_{j-1})\, dr \right\|_{X_0}.$$

Finally, applications of (7.7b), of (7.10c), with $r = m$ and $p = s - \gamma + m$, and of (7.14), with $\sigma = s$, yield

$$\begin{aligned}
\|\mathbf{F}'(u_{j-1})[\mathbf{I} - \mathbf{S}_{t_j}]\mathbf{G}_{h_j}(u_{j-1})\mathbf{F}(u_{j-1})\|_{X_0} &\leq\\
M\|[\mathbf{I} - \mathbf{S}_{t_j}]\mathbf{G}_{h_j}(u_{j-1})\mathbf{F}(u_{j-1})\|_{Y_0} &\leq\\
M^2 t_j^{-(s-\gamma)} \|\mathbf{G}_{h_j}(u_{j-1})\mathbf{F}(u_{j-1})\|_{Y_{s-\gamma}} &\leq M^3 t_j^{-(s-\gamma)} \|\mathbf{F}(u_{j-1})\|_{X_s}.
\end{aligned}$$

We now estimate, separately, by use of the induction hypothesis. Applications of (7.26c) to the first of the three derived inequalities, of (7.26f) to the second, and of (7.26b) to the third, all with $k = j - 1$ according to the induction hypothesis, give

$$\|\mathbf{F}(u_j)\|_{X_0} \leq 2M^9 \rho^{-2(\lambda - \beta\gamma)\beta^{j-1}} + M^4 \rho^{-\theta(s-\gamma)\beta^j} \rho^{(s-\lambda)\beta^j}. \qquad (7.28)$$

When the inequalities,

$$2(\lambda - \beta\gamma) \geq \lambda\beta + \lambda_0, \qquad \theta(s - \gamma) - (s - \lambda) \geq \lambda + s_0,$$

which follow from (7.20b, 7.20c), respectively, are applied to (7.28), we obtain:

$$\|\mathbf{F}(u_j)\|_{X_0} \leq 2M^9 \rho^{-\lambda_0} \rho^{-\lambda\beta^j} + M^4 \rho^{-s_0} \rho^{-\lambda\beta^j}. \qquad (7.29)$$

This yields (7.26a), with $k = j$, upon use of (7.21b, 7.21c). Notice that $\rho \geq 1$ was used in deriving (7.29). Also, the case $j = 0$ in (7.26a) is immediate from the hypothesis, (7.15).

Inequality (7.26b), for $k = j > 0$, follows from the induction hypothesis: $\|u_j - u_0\|_{Y_s} \leq \rho^{(s-\lambda)\beta^{j+1}}$, and from (7.12), with $a = s$, $v_0 = u_0$, if account is taken of the induction hypothesis: $\|u_j - u_0\|_{Y_\gamma} \leq 1$, and of (7.19). Again, note that $\rho \geq 1$. Inequality (7.26b), for $j = 0$, is immediate from the hypothesis (7.16).

To establish (7.26c), for $k = j \geq 0$, we begin with the interpolation inequality (7.11), taking $q = \gamma/s$, $r_1 = 0$, and $r_2 = s$, and obtain

$$\|\mathbf{F}(u_j)\|_{X_\gamma} \leq M\|\mathbf{F}(u_j)\|_{X_0}^{1-\frac{\gamma}{s}}\|\mathbf{F}(u_j)\|_{X_s}^{\gamma/s}. \tag{7.30}$$

Applying the residual estimates which we have just derived, i.e., (7.26a, b) for $k = j$, we obtain from (7.30) the estimate,

$$\begin{aligned}\|\mathbf{F}(u_j)\|_{X_\gamma} &\leq M^2\rho^{-\lambda\beta^j\left(1-\frac{\gamma}{s}\right)}\rho^{(s-\lambda)\beta^{j+1}(v/s)} \\ &= M^2\rho^{-\beta^j\left(\lambda-\beta\gamma+\left(\frac{\lambda\gamma}{s}\right)(\beta-1)\right)} \leq M^2\rho^{-(\lambda-\beta\gamma)\beta^j},\end{aligned}$$

since $\rho,\ M \geq 1$. This is just (7.26c) for $k = j$.

For (7.26d), we use the definition (7.18) in conjunction with (7.10b), as applied to Y_σ for $p = s$, and $r = s - \gamma$. When this is coupled with (7.14), for $\sigma = s$, we obtain

$$\|u_{j+1} - u_j\|_{Y_s} \leq M^2(t_{j+1})^\gamma\|\mathbf{F}(u_j)\|_{X_s},$$

and this, in turn, is estimated by (7.17) and (7.26b), for $k = j$. Altogether, we have

$$\|u_{j+1} - u_j\|_{Y_s} \leq M^3\rho^{\theta\gamma\beta^{j+1}}\rho^{(s-\lambda)\beta^{j+1}},$$

and the right hand side of this expression is bounded by $M^3\rho^{-s_0}\rho^{(s-\lambda)\beta^{j+2}}$, upon use of the inequality

$$s - \lambda + \theta\gamma \leq (s - \lambda)\beta - s_0.$$

The latter follows from (7.20c). If we now apply (7.21c), we obtain (7.26d), with $k = j \geq 0$.

We now verify (7.26e) for $k = j > 0$. Indeed, we have

$$\begin{aligned}\|u_{j+1} - u_0\|_{Y_s} &\leq \|u_{j+1} - u_j\|_{Y_s} + \sum_{k=1}^{j}\|u_k - u_{k-1}\|_{Y_s} \\ &\leq \tfrac{1}{2}\rho^{(s-\lambda)\beta^{j+2}} + \tfrac{1}{2}\sum_{k=1}^{j}\rho^{(s-\lambda)\beta^{k+1}},\end{aligned} \tag{7.31}$$

upon use of (7.26d) and the induction hypothesis. The second sum is estimated, via the lower sums, by the elementary technique of dominating it by the integral $I(j+1)$, where

$$I(b) = \tfrac{1}{2} \int_1^b \rho^{(s-\lambda)\beta^{x+1}} \, dx.$$

An evident comparison of areas yields

$$I(j+1) \geq \tfrac{1}{2} \sum_{k=1}^{j} \rho^{(s-\lambda)\beta^{k+1}}, \tag{7.32}$$

since the function,

$$f(x) = \rho^{(s-\lambda)\beta^{x+1}},$$

is (strictly) increasing on $[1, j+1]$. In order to estimate $I(j+1)$, we compute

$$f'(x) = \rho^{(s-\lambda)\beta^{x+1}} \beta^{x+1} \ln(\rho^{(s-\lambda)}) \ln \beta,$$

so that, for $1 \leq x \leq j+1$,

$$f(x) \leq f'(x)/(\beta^2 (s-\lambda) \ln \rho \ln \beta). \tag{7.33}$$

If (7.33) is integrated over the interval $[1, j+1]$, we have

$$I(j+1) = \tfrac{1}{2} \int_1^{j+1} f(x) \, dx \leq \tfrac{1}{2} \int_1^{j+1} f'(x) \, dx \leq \tfrac{1}{2} f(j+1) = \tfrac{1}{2} \rho^{(s-\lambda)\beta^{j+2}}. \tag{7.34}$$

Here we have used the inequality,

$$\beta^2 (s-\lambda) \ln \rho \ln \beta \geq 1, \tag{7.35}$$

which follows the observations that $\beta \mapsto \beta^2 \ln \beta/(\beta - 1)$ increases from 1 on $[1, 2]$ and $\rho \geq e$ (cf. (7.21a)). Combining (7.31, 7.32, 7.34) gives (7.26e) for $k = j$.

The verification of (7.26f) reduces to the inequality,

$$\|u_{j+1} - u_j\|_{Y_0} \leq M^2 \|\mathbf{F}(u_j)\|_{X_\gamma}, \tag{7.36}$$

as is seen by comparison of (7.26c) and (7.26f). However, if we begin with (7.18), and apply (7.10b), in terms of the spaces, $\{Y_\sigma\}$, with $r = p = 0$, and (7.14), with $\sigma = \gamma$, we obtain (7.36). Inequality (7.26g) is verified similarly to (7.26e), by making use of the decreasing function,

$$g(x) = \rho^{-(\lambda - \beta\gamma)\beta^x}, \qquad 0 \leq x \leq j.$$

With this substitution, we obtain the estimate,

$$\begin{aligned} M^{-4}\|u_{j+1}-u_0\|_{Y_0} &\le \sum_{k=0}^{j}\rho^{-(\lambda-\beta\gamma)\beta^k} \le \rho^{-(\lambda-\beta\gamma)} + \int_0^j g(x)\,dx \\ &\le \rho^{-(\lambda-\beta\gamma)} + [(\lambda-\beta\gamma)\ln\rho\ln\beta]^{-1}\int_0^j(-g'(x))\,dx \\ &\le \{1+[(\lambda-\beta\gamma)\ln\rho\ln\beta]^{-1}\}\rho^{-(\lambda-\beta\gamma)} \\ &\le 2M^{-4}\rho^{-(\lambda-\beta\gamma)} \\ &\le M^{-4}\eta, \end{aligned} \tag{7.37}$$

upon use of (7.21a, d) and (7.20b).

We next prove (7.26h). We begin with an interpolation inequality similar to (7.30), for δ satisfying $\gamma \le \delta \le \mu$:

$$\|u_{j+1}-u_j\|_{Y_\delta} \le M\|u_{j+1}-u_j\|_{Y_0}^{1-\frac{\delta}{s}}\,\|u_{j+1}-u_j\|_{Y_s}^{\delta/s},$$

and estimate the right hand side terms by (7.26f) and (7.26d). This gives

$$\|u_{j+1}-u_j\|_{Y_\delta} \le M^5\rho^{-[(\lambda-\beta\gamma)(1-\frac{\delta}{s})-\beta^2(s-\lambda)(\frac{\delta}{s})]\beta^j}.$$

If the inequality,

$$(\delta/s)(\beta^2\lambda-\lambda+\beta\gamma)\ge 0,$$

is applied to the preceding inequality, we obtain

$$\|u_{j+1}-u_j\|_{Y_\delta} \le M^5\rho^{-(\lambda-\beta\gamma-\beta^2\delta)\beta^j} \le M^5\rho^{-\lambda_0\beta^j},$$

where we have used (7.20a, b). This establishes (7.26h), with $k=j\ge 0$.

It remains to establish (7.26i). By an estimate similar to (6.2.31), we obtain:

$$M^{-5}\|u_{j+1}-u_0\|_{Y_\delta} \le \sum_{k=0}^{j}\rho^{-\lambda_0\beta^k} \le \rho^{-\lambda_0} + [\lambda_0\ln\rho\ln\beta]^{-1}\rho^{-\lambda_0},$$

and inequality (7.26i) follows from (7.21b, e).

We now prove that $\{u_k\}$ is a Cauchy sequence in Y_0 and in Y_μ, satisfying the estimates,

$$\|u_{k+m}-u_k\|_{Y_0} \le M^4\rho^{-(\lambda-\beta\gamma)\beta^{k-1}}/\beta^{k-1}, \tag{7.38a}$$

$$\|u_{k+m}-u_k\|_{Y_\mu} \le M^5\rho^{-\lambda_0\beta^{k-1}}/\beta^{k-1}, \tag{7.38b}$$

for k, $m\ge 1$. An application of (7.26f, 7.26h) gives

$$\|u_{k+m}-u_k\|_{Y_0} \le M^4\sum_{j=k}^{k+m-1}\rho^{-(\lambda-\beta\gamma)\beta^j},$$

and

$$\|u_{k+m} - u_k\|_{Y_\mu} \le M^5 \sum_{j=k}^{k+m-1} \rho^{-\lambda_0 \beta^j}.$$

Estimation of these sums by improper integrals gives

$$\|u_{k+m} - u_k\|_{Y_0} \le M^4 \int_{k-1}^{\infty} \rho^{-(\lambda-\beta\gamma)\beta^x}\, dx, \tag{7.39a}$$

and

$$\|u_{k+m} - u_k\|_{Y_0} \le M^5 \int_{k-1}^{\infty} \rho^{-\lambda_0\beta^x}\, dx. \tag{7.39b}$$

Both integrals are of the form $\int_{k-1}^{\infty} \rho^{-\delta\beta^x}\, dx$, $\delta \ge \lambda_0$, which can be estimated as

$$\begin{aligned}\int_{k-1}^{\infty} \rho^{-\delta\beta^x}\, dx &\le [\beta^{k-1}\delta \ln\rho \ln\beta]^{-1} \int_{k-1}^{\infty} \left[\tfrac{d}{dx}(-\rho^{-\delta\beta^x})\right] dx \\ &\le [\beta^{k-1}\delta \ln\rho \ln\beta]^{-1} \rho^{-\delta\beta^{k-1}} \le \rho^{-\delta\beta^{k-1}}/\beta^{k-1}. \end{aligned} \tag{7.40}$$

Here we have used (7.21e).

The estimates (7.2.32a, b) follow from (7.39a, b) and (6.2.34). Since $\{u_k\}$ is a Cauchy sequence in Y_0 and Y_μ, and since the injection $Y_\mu \mapsto Y_0$ is continuous, it follows that there is an element $u \in B_{\eta,0} \cap B_{1,\mu}$, such that $u_k \to u$ in both topologies. Inequalities (7.23, 7.24) follow from (7.2.32a, b) by letting $m \to \infty$. The fact that u is a root, i.e., $\mathbf{F}(u) = 0$, follows routinely from (7.26a) and the continuity of $\mathbf{F}$ on $B_{\eta,0}$.

Remark 7.2.3. In the reference [65], there is a detailed discussion of analytical realizations of the smoothing in terms of high frequency filtering.

References

1. R.A. Adams: *Sobolev Spaces.* Academic Press, New York 1975
2. F. Alabau: A method for proving uniqueness theorems for the stationary semiconductor device and electrochemistry equations. Nonlinear Anal. **18** (1992) 861–872
3. F. Alabau: Comportement de la corbe caractéristique potentiel appliqué-courant d'une diode en polarisation inverse et directe. C.R. Acad. Sci. Série I **314** (1992) 881–886
4. M.G. Ancona and G.J. Iafrate: Quantum correction to the the equation of state of an electron gas in a semiconductor. Phys. Rev. B**39** (1989) 9536–9540
5. M.G. Ancona and H.F. Tiersten: Macroscopic physics of the silicon inversion layer. Phys. Rev. B**35** (1987) 7959–7965
6. J.-P. Aubin: *Approximation of Elliptic Boundary-Value Problems.* Wiley, New York 1972
7. A. Azzam and E. Kreyszig: On solutions of elliptic equations satisfying mixed boundary conditions. SIAM J. Math. Anal. **13** (1982) 254–262
8. I. Babuška and A.K. Aziz: Survey lectures on the mathematical foundations of the finite element method. In: A.K. Aziz (ed.) *The Mathematical Foundations of the Finite Element Method with Applications to Partial Differential Equations.* Academic Press, New York 1972, pp. 5–359
9. G. Baccarani and M.R. Wordeman: An investigation of steady-state velocity overshoot effects in Si and GaAs devices. Solid State Electr. **28** (1985) 407–416
10. R.E. Bank, J.W. Jerome, and D.J. Rose: Analytical and numerical aspects of semiconductor device modeling. In: R. Glowinski and J. Lions (eds.) *Proc. Fifth International Conference on Computing Methods in Applied Science and Engineering.* North Holland, Amsterdam 1982, pp. 593–597
11. R.E. Bank (ed.) *Computational Aspects of VLSI Design with an Emphasis on Semiconductor Device Simulation. Lectures in Applied Mathematics*, vol. 25. American Mathematical Society, Providence, R.I. 1990
12. R.E. Bank, R. Bulirsch, and K. Merten (eds.) *Mathematical Modeling and Simulation of Electrical Circuits and Semiconductor Devices.* Birkhäuser, Basel 1990
13. R.E. Bank, D.J. Rose, and W. Fichtner: Numerical methods for semiconductor device simulation. IEEE Trans. Electron Devices ED-**30** (1983) 1031–1041
14. J. Bergh and J.Löfström: *Interpolation Spaces.* Springer, Berlin Heidelberg New York 1986
15. F.J. Blatt: *Physics of Electric Conduction in Solids.* McGraw Hill, New York 1968
16. K. Blotekjaer: Transport equations for electrons in two-valley semiconductors. IEEE Trans. Electron Devices ED-**17** (1970) 38–47
17. S.C. Brenner and L.R. Scott: *The Mathematical Theory of Finite Element Methods.* Springer, New York 1994

18. H. Brézis: Équations et inéquations non linéaires dans les espaces vectoriels en dualité. Ann. Inst. Fourier (Grenoble) **18** (1968) 115–175
19. F. Brezzi: On the existence, uniqueness, and approximation of saddle point problems arising from Lagrangian multipliers. R.A.I.R.O., Anal. Numér. **12** (1974) 129–151
20. F. Brezzi, A. Capello, and L. Gastaldi: A singular perturbation analysis of reverse biased semiconductor diodes. SIAM J. Math. Anal. **20** (1989) 372–387
21. S. Busenberg, W. Fang, and K. Ito: Modeling and analysis for laser-beam-induced current images in semiconductors. SIAM J. Appl. Math. **53** (1993) 187–204
22. E.M. Buturla, P.E. Cottrell, B.M. Grossman, and K.A. Salsburg: Finite-element analysis of semiconductor devices: the FIELDAY program. IBM J. Res. Dev. **25** (1981) 218–231
23. C. Cercignani: *The Boltzmann Equation and its Application*. Springer, New York 1987
24. C. Cercignani, R. Illner, and M. Pulvirenti: *The Mathematical Theory of Dilute Gases*. Springer, New York 1994
25. T. Chan and K. Jackson: Nonlinearly preconditioned Krylov subspace methods for discrete Newton algorithms. SIAM J. Sci. Stat. Comput. **5** (1984) 533–542
26. Z. Chen, B. Cockburn, C. Gardner, and J. Jerome: Quantum hydrodynamic simulation of hysteresis in the resonant tunneling diode. J. Comp. Phys. **117** (1995) 274-280
27. Z. Chen, B. Cockburn, J. Jerome, and C.-W. Shu: Mixed-RKDG finite element methods for the 2-D hydrodynamic model for semiconductor device simulation. VLSI DESIGN **3** (1995) 145–158
28. D. Chen, E. Kan, U. Ravaioli, C. Shu, and R. Dutton: An improved energy transport model including non-parabolic and non-Maxwellian distribution effects. IEEE Elec. Dev. Lett. **13** (1992) 26–28
29. P.G. Ciarlet and P.-A. Raviart: Maximum principle and uniform convergence for the finite element method. Computer Methods in Applied Mechanics and Engineering **2** (1973) 17–31
30. W.M. Coughran and J.W. Jerome: Modular algorithms for transient semiconductor device simulation, Part I: Analysis of the outer iteration. In: R.E. Bank (ed.) *Computational Aspects of VLSI Design with an Emphasis on Semiconductor Device Simulation. Lectures in Applied Mathematics*, vol. 25. American Mathematical Society, Providence, R.I. 1990, pp. 107–149
31. P. Degond, F. Guyot-Delaurens, F.J. Mustieles, and F. Nier: Semiconductor modelling via the Boltzmann equation. In: R.E. Bank, R. Bulirsch, and K. Merten (eds). *Mathematical Modelling and Simulation of Electrical Circuits and Semiconductor Devices*. Birkhäuser, Basel 1990, pp. 153–167
32. P. Degond and B. Niclot: Numerical analysis of the weighted particle method applied to the semiconductor Boltzmann equation. Numer. Math. **55** (1989) 599–618
33. I. Ekeland and R. Temam: *Convex Analysis and Variational Problems*. North-Holland, Amsterdam New York 1976
34. W.L. Engl and H. Dirks: Numerical device simulation guided by physical approaches. In: B.T. Browne and J.J.H. Miller (eds.) *Numerical Analysis of Semiconductor Devices*. Boole Press, Dublin 1979, pp. 65–93
35. E. Fatemi: Linear Analysis of the Hydrodynamic Model. Numer. Functional Anal. Optim. **16** (1995) 303–314
36. E. Fatemi, C. Gardner, J. Jerome, S. Osher, and D. Rose: Simulation of a steady-state electron shock wave in a submicron semiconductor device using high order upwind methods. In: K. Hess, J. P. Leburton, and U. Ravaioli

(eds.) *Computational Electronics*. Kluwer Academic Publishers, Boston Dordrect 1991, pp. 27–32

37. E. Fatemi, J. Jerome, and S. Osher: Solution of the hydrodynamic device model using high-order nonoscillatory shock capturing algorithms. IEEE Trans. Computer-Aided Design of Integrated Circuits and Systems CAD-**10** (1991) 232–244
38. W. Fichtner: Physics of VLSI processing and process simulation. In: D. Kahng (ed.) *Silicon Integrated Circuits, Part C*. Academic Press, Orlando 1985, pp. 119–336
39. M.V. Fischetti and S.E. Laux: Monte Carlo analysis of electron transport on small semiconductor devices including band-structure and space-charge effects. Phys. Rev. B**38** (1988) 9721–9745
40. A. Friedman: *Advanced Calculus*. Holt, Rinehart, Winston, New York 1971
41. I.M. Gamba: Stationary transonic solutions for a one-dimensional hydrodynamic model for semiconductors. Communications in P.D.E. **17** (1992) 553–577
42. I.M. Gamba: Asymptotic behavior at the boundary of a semiconductor device in two space dimensions. Annali Mat. Pura Appl. **163** (1993) 43–91
43. I.M. Gamba: Viscosity approximating solutions to ODE systems that admit shocks, and their limits. Advances in Appl. Math. **15** (1994) 129–182
44. C.L. Gardner: Numerical simulation of a steady-state electron shock wave in a submicrometer semiconductor device. IEEE Trans. Electron Devices ED-**38** (1991) 392–398
45. C.L. Gardner: Hydrodynamic and Monte-Carlo simulation of an electron shock wave in a $1-\mu m$ n^+-n-n^+ diode. IEEE Trans. Electron Devices ED-**40** (1993) 455–457
46. C.L. Gardner: The quantum hydrodynamic model for semiconductor devices. SIAM J. Appl. Math. **54** (1994) 409–427
47. C.L. Gardner, J.W. Jerome, and D.J. Rose: Numerical methods for the hydrodynamic device model: Subsonic flow. IEEE Trans. Computer-Aided Design of Integrated Circuits and Systems CAD-**8** (1989) 501–507
48. D. Gilbarg and N. Trudinger: *Elliptic Partial Differential Equations of Second Order*. Springer, New York 1977
49. A. Gnudi, F. Odeh, and M. Rudan: Investigation of non-local transport phenomena in small semiconductor devices. European Trans. on Telecommunications and Related Technologies **1**(**3**) (1990) 307–312 (77-82)
50. S. Goldstein: *Lectures on Fluid Mechanics*. Wiley Interscience, London New York 1957
51. T.N.E. Greville: Interpolation by generalized spline functions. Technical Report 476, Mathematics Research Center, University of Wisconsin 1964
52. P. Grisvard: *Elliptic Problems in Nonsmooth Domains. Monographs and Studies in Mathematics*, vol. 24. Pitman, London 1985
53. H.L. Grubin and J.P. Kreskovsky: Quantum moment balance equations and resonant tunneling structures. Solid State Electronics **32** (1989) 1071–1075
54. H.K. Gummel: A self-consistent iterative scheme for one-dimensional steady state transistor calculations. IEEE Trans. Electron Devices ED-**11** (1964) 455–465
55. B. Gustafsson and A. Sundstrom: Incompletely parabolic systems in fluid dynamics. SIAM J. Appl. Math. **35** (1978) 343–357
56. W. Hänsch and M. Miura-Mattausch: The hot-electron problem in small semiconductor devices. J. Appl. Physics **60** (1986) 650–656
57. K. Hess, J.P. Leburton, and U. Ravaioli (eds.) *Computational Electronics*. Kluwer Academic Publishers, Boston Dordrecht 1991

58. K. Hess: *Advanced Theory of Semiconductor Devices.* Prentice-Hall, Englewood Cliffs, N.J. 1988
59. R.W. Hockney and J.W. Eastwood: Computer Simulation Using Particles McGraw -Hill, New York 1981
60. K. Huang: *Statistical Mechanics.* John Wiley, New York 1987
61. J.W. Jerome: *On the L_2 n-Width of Certain Classes of Functions of Several Variables.* PhD thesis, Purdue University, Lafayette, Indiana 1966
62. J.W. Jerome: Asymptotic estimates of the L_2 n-width. J. Math. Anal. Appl. **22** (1968) 449–464
63. J.W. Jerome: Uniform approximation by certain generalized spline functions. J. Approximation Theory **7** (1973) 143–154
64. J.W. Jerome: *Approximation of Nonlinear Evolution Systems.* Academic Press, New York 1983
65. J.W. Jerome: An adaptive Newton algorithm based on numerical inversion: Regularization as postconditioner. Numer. Math. **47** (1985) 123–138
66. J.W. Jerome: Approximate Newton methods and homotopy for stationary operator equations. Constr. Approx. **1** (1985) 271–285
67. J.W. Jerome: Consistency of semiconductor modelling: An existence/stability analysis for the stationary Van Roosbroeck system. SIAM J. Appl. Math. **45** (1985) 565–590
68. J.W. Jerome: The role of semiconductor device diameter and energy-band bending in convergence of Picard iteration for Gummel's map. IEEE Trans. Electron Devces ED-**32** (1985) 2045–2051
69. J.W. Jerome: Evolution systems in semiconductor device modeling: A cyclic uncoupled line analysis for the Gummel map. Math. Methods in the Appl. Sc. **9** (1987) 455–492
70. J.W. Jerome: Newton's method for gradient equations based upon the fixed point map. Numer. Math. **55** (1989) 619–632
71. J.W. Jerome: Numerical approximation of PDE system fixed-point maps via Newton's method. J. Comp. Appl. Math. **38** (1991) 211–230
72. J.W. Jerome: The mathematical study and approximation of semiconductor models. In: J. Gilbert and D. Kershaw (eds.) *Advances in Numerical Analysis: Large Scale Matrix Problems and the Numerical Solution of Partial Differential Equations.* Oxford University Press 1994, pp. 157–204
73. J.W. Jerome: An asymptotically linear fixed point extension of the inf-sup theory of Galerkin approximation. Numer. Functional Anal. Optim. **16** (1995) 345–361
74. J.W. Jerome and T. Kerkhoven: A finite element approximation theory for the drift-diffusion semiconductor model. SIAM J. Numer. Anal. **28** (1991) 403–422
75. J.W. Jerome and C.-W. Shu: Energy models for one-carrier transport in semiconductor devices. In: W.M. Coughran, J. Cole, P. Lloyd, and J.K. White (eds.) *Semiconductors, Part II. IMA Volumes in Mathematics and its Applications*, vol. 59. Springer, New York 1994, pp. 185–207
76. J.W. Jerome and C.-W. Shu: Transport effects and characteristic modes in the modeling and simulation of submicron devices. IEEE Trans. Computer-Aided Design of Integrated Circuits and Systems CAD-**14** (1995) 917–923
77. J.W. Jerome and C.-W. Shu: Energy transport systems for semiconductors: Analysis and simulation. In: *Proceedings, First World Congress of Nonlinear Analysts.* Walter de Gruyter, Berlin 1995
78. J.W. Jerome and C.-W. Shu: The response of the hydrodynamic model to heat conduction, mobility, and relaxation expressions. VLSI DESIGN **3** (1995) 131–143

79. J.W. Jerome and L.L. Schumaker: Local support bases for a class of spline functions. J. Approximation Theory **16** (1976) 16–27
80. Claes Johnson: *Numerical Solutions of Partial Differential Equations by the Finite Element Method.* Cambridge University Press 1987
81. R.B. Kellogg: Higher order singularities for interface problems. In: A.K. Aziz (ed.) *The Mathematical Foundations of the Finite Element Method with Applications to Partial Differential Equations.* Academic Press, New York 1972, pp. 589–602
82. T. Kerkhoven: *Coupled and Decoupled Algorithms for Semiconductor Simulation.* PhD thesis, Yale University, New Haven, Connecticut 1985
83. T. Kerkhoven: On the effectiveness of Gummel's method. SIAM J. Sci. & Stat. Comp. **9** (1988) 48–60
84. T. Kerkhoven and J.W. Jerome: L_∞ stability of finite element approximations to elliptic gradient equations. Numer. Math. **57** (1990) 561–575
85. T. Kerkhoven and Y. Saad: On acceleration methods for coupled nonlinear elliptic systems. Numer. Math. **60** (1992) 525–548
86. M. Kern: Non-zero space charge approximation of thyristors. Math. Nachr. **157** (1992) 169–183
87. C. Kittel: *Introduction to Solid State Physics.* John Wiley, New York 1976
88. N.C. Kluksdahl, A.M. Kriman, D.K. Ferry, and C. Ringhofer: Self-consistent study of the resonant tunneling diode. Phys. Rev. B**39** (1989) 7720–7735
89. A.N. Kolmogorov: Über die beste Annähergung von Funktionen einer gegebenen Funktionklasse. Ann. Math. **37** (1936) 107–111
90. M.A. Krasnosel'skii, G.M. Vainikko, P.P. Zabreiko, Ya.B. Rititskii, and V.Ya. Stetsenko: *Approximate Solution of Operator Equations.* Wolters-Noordhoff, Groningen 1972
91. M.A. Krasnosel'skii, P.P. Zabreiko, E.I. Pustylnik, and P.E. Sobolevskii: *Integral Operators in Spaces of Summable Functions.* Wolters-Noordhoff, Groningen 1976
92. A. Kufner: Einige Eigenschaften der Sobolevschen mit Belegungsfunktion. Czechoslovakian J. Math. **15** (1965) 597–619
93. L.D. Landau and E.M. Lifshitz: *Fluid Mechanics*, 2nd. ed. Course of Theoretical Physics, vol. 6. Pergamon Press, Oxford 1987
94. P.D. Lax and C.D. Levermore: The zero dispersion limit for the Korteweg-deVries equation. Proc. Nat. Acad. Sci. USA **76** (1979) 3602–3606
95. Y.L. Le Coz: *Semiconductor Device Simulation: A Spectral Method for Solution of the Boltzmann Transport Equation.* PhD thesis, Massachusetts Inst. Tech. Cambridge, Mass. 1988
96. R.L. Liboff: *Introduction to the Theory of Kinetic Equations.* John Wiley, New York 1969
97. J.L. Lions and E. Magenes: *Non-Homogeneous Boundary Value Problems and Applications*, vol. I. Springer, Berlin Heidelberg New York 1972
98. G.G. Lorentz: *Approximation Theory.* Holt, New York 1966
99. M. Lundstrom: *Fundamentals of Carrier Transport.* Addison -Wesley, Reading, MA 1990
100. P.A. Markowich: *The Stationary Semiconductor Device Equations.* Springer, Vienna and New York 1986
101. P.A. Markowich, C.A. Ringhofer, and C. Schmeiser: *Semiconductor Equations.* Springer, Vienna 1990
102. S.G. Mikhlin: *The Problem of the Minimum of a Quadratic Functional.* Holden-Day, San Francisco 1965
103. M.S. Mock: On equations describing steady-state carrier distributions in a semiconductor device. Comm. Pure Appl. Math. **25** (1972) 781–792

104. M.S. Mock: *Analysis of Mathematical Models of Semiconductor Devices.* Boole Press, Dublin 1983
105. J. Moser: A rapidly convergent iteration method and nonlinear partial differential equations, I. Ann. Scuola Norm. Sup. Pisa Cl. Sci. **XX** (1966) 265–315
106. M.K.V. Murthy and G. Stampacchia: A variational inequality with mixed boundary conditions. Israel J. Math. **13** (1972) 188–224
107. B. Niclot, P. Degond, and F. Poupaud: Deterministic particle simulations of the Boltzmann transport equation of semiconductors. J. Comp. Phys. **78** (1988) 313–339
108. L. Nirenberg: *Topics in Nonlinear Functional Analysis.* Courant Institute of Mathematical Sciences, New York University 1973
109. J.P. Nougier, J. Vaissiere, D. Gasquet, J. Zimmermann, and E. Constant: Determination of the transient regime in semiconductor devices using relaxation time approximations. J. Appl. Phys. **52** (1981) 825–832
110. J.M. Ortega and W.C. Rheinboldt: *Iterative Solution of Nonlinear Equations in Several Variables.* Academic Press, New York 1970
111. A. Pinkus: *n-Widths in Approximation Theory.* Springer, Berlin Heidelberg New York 1985
112. F. Riesz and B. Sz.-Nagy: *Functional Analysis.* Ungar, New York 1955
113. W. Van Roosbroeck: Theory of flow of electrons and holes in germanium and other semiconductors. Bell System Tech. J. **29** (1950) 560–607
114. I. Rubinstein: Multiple steady states in one-dimensional electrodiffusion with local electroneutrality. SIAM J. Appl. Math. **47** (1987) 1076–1093
115. I. Rubinstein: *Electro-Diffusion of Ions. SIAM Studies in Applied Mathematics.* SIAM, Philadelphia 1990
116. M. Rudan and F. Odeh: Multi-dimensional discretization scheme for the hydrodynamic model of semiconductor devices. COMPEL **5** (1986) 149–183
117. D.L. Scharfetter and H.K. Gummel: Large signal analysis of a silicon Read diode oscillator. IEEE Trans. Electron Devices ED-**16** (1969) 64–77
118. Y. Saad and M.H. Schultz: GMRES: A generalized minimal residual method for solving nonsymmetric linear systems. SIAM J. Sci. Stat. Comput. **7** (1986) 856–869
119. T. Seidman: Steady state solutions of diffusion reaction systems with electrostatic convection. Nonlinear Anal. **4** (1980) 623–637
120. S. Selberherr: *Analysis and Simulation of Semiconductor Devices.* Springer, New York 1984
121. N. Shigyo, S. Onga, and R. Dang: A three-dimensional mos device simulator. In: J.J.H. Miller (ed.) *New Problems and New Solutions for Device and Process Modeling.* Boole Press, Dublin 1985, pp. 138–149
122. W. Shockley: *Electrons and Holes in Semiconductors.* Van Nostrand, Princeton, N.J. 1950
123. C.-W. Shu and S.J. Osher: Efficient implementation of essentially non-oscillatory shock capturing schemes, II. J. Comp. Phys. **83** (1989) 32–78
124. J.C. Srikwerda: Initial boundary value problems for incompletely parabolic systems. Comm. Pure Appl. Math. **30** (1977) 797–822
125. H. Steinrueck and F. Odeh: The Wigner function for thermal equilibrium. ZAMP **42** (1991) 470–487
126. M.A. Stettler, M.A. Alam, and M.S. Lundstrom: A critical examination of the assumptions underlying macroscopic transport equations for silicon devices. IEEE Trans. Electron Devices ED-**40** (1993) 733–740
127. G.W. Strang: Approximation in the finite element method. Numer. Math. **19** (1972) 81–98

128. G.W. Strang and G. Fix: *An Analysis of the Finite Element Method.* Prentice-Hall, Englewood Cliffs, N.J. 1973
129. B.G. Streetman: *Solid State Electronic Devices.* Prentice-Hall, Englewood Cliffs, NJ 1980
130. S.M. Sze: *Physics of Semiconductor Devices.* John Wiley, New York 1981
131. G.-L. Tan, X.-L. Yuan, Q.-M. Zhang, W.H. Tu, and A.-J. Shey: Two-dimensional semiconductor device analysis based on new finite-element discretization employing the S-G scheme. IEEE Trans. Computer-Aided Design of Integrated Circuits and Systems CAD-**8** (1989) 468–478
132. R. Tapia: The differentiation and integration of nonlinear operators. In: L.B. Rall (ed.) *Nonlinear Functional Analysis and Applications.* Academic Press, New York 1971, pp. 45–101
133. T. Toyabe, H. Masuda, Y. Aoki, H. Shukuri, and T. Hagiwara: Three-dimensional device simulator CADDETH with highly convergent matrix solution algorithms. IEEE Trans. Electron Devices ED-**32** (1985) 2038–2045
134. R.S. Varga: *Matrix Iterative Analysis.* Prentice-Hall, Englewood Cliffs, N.J. 1962
135. T. Wang and K. Hess: Calculation of the electron velocity distribution in high electron mobility transistors using an ensemble Monte Carlo method. J. Appl. Phys. **57** (1985) 5336–5339
136. J. White and A. Sangiovanni-Vincentelli: *Relaxation Techniques for the Simulation of VLSI Circuits.* Kluwer Academic Publishing Co., Boston, Dordrecht 1987
137. N.M. Wigley: Mixed boundary value problems in plane domains with corners. Math. Z. **115** (1970) 33–52
138. E. Wigner: On the quantum correction for thermodynamic equilibrium. Phys. Rev. **40** (1932) 749–759
139. B. Zhang: Convergence of the Godunov scheme for a simplified one-dimensional hydrodynamic model for semiconductor devices. Commun. Math. Phys. **157** (1993) 1–22
140. B. Zhang and J.W. Jerome: On a steady-state quantum hydrodynamic model for semiconductors. Nonlinear Analysis, in press

Index

Springer-Verlag and the Environment

We at Springer-Verlag firmly believe that an international science publisher has a special obligation to the environment, and our corporate policies consistently reflect this conviction.

We also expect our business partners – paper mills, printers, packaging manufacturers, etc. – to commit themselves to using environmentally friendly materials and production processes.

The paper in this book is made from low- or no-chlorine pulp and is acid free, in conformance with international standards for paper permanency.